SPACE FLIGHT

THE RECORDS

GUINNESS BOOKS

SPACE FLIGHT

THE RECORDS

Tim Furniss

Dedication

To Christine,
with Love and Thanks

Editor: Beatrice Frei
Design and layout: Stonecastle Graphics

Published in Great Britain by Guinness Superlatives Ltd,
2 Cecil Court, London Road, Enfield, Middlesex

Typeset in Times and Helvetica
Phototypeset by Input Typesetting Ltd, London, SW19 8DR
Printed and bound in Great Britain by
Hazell Watson & Viney Ltd, Member of the BPCC
Group, Aylesbury, Buckinghamshire.

British Library Cataloguing in Publication Data

Furniss, Tim
Guinness space: the records.
1. Manned space flight—History
I. Title
629.45′009 TL788.5

ISBN 0–85112–451–8 Pbk

Contents

Introduction

During the sixties when I was at school, whenever an American manned space shot was imminent, I took along my transistor radio to listen to the launch. I well remember the dreaded Tuesday afternoon's double-economics period on 25 March 1965 being thankfully interrupted by live coverage of the launch of Gemini 3 from my radio! In those days, manned space flight received a lot of coverage in the press and on television and radio.

Today, twenty years later, a shuttle launch may still just rate news coverage but it lacks, not surprisingly, factual information, other than perhaps reference to the number of crew and the duration of the mission. Thus, many significant – and not so significant – records, firsts, facts and feats are chalked up almost dismissively these days, and remain unrecorded.

Space Flight – The Records is intended to fill this gap. Its publication is timely, as we are approaching the twenty-fifth anniversary of Yuri Gagarin's flight into space on 12 April 1961. It is, however, purely a facts and figures book which does not set out to answer the question, 'Why space?'; nor have I included details of experiments and cargoes carried into space.

I have never ceased to be amazed (and while preparing this book, frustrated nearly to the point of tearing my hair out) at the variations – however slight – in data from seemingly official sources, such as NASA documents and Novosti press releases. This is particularly true of spacecraft weights, EVA times and in some cases flight times which are, incidentally, given to the second, where available. This fact has had to be accepted as the occupational hazard of putting this book together.

I would especially like to thank three people who have helped me a great deal in providing some elusive data for this book. They are Rex Hall, Neville Kidger and David Shayler who all belong to the British Interplanetary Society and who are expert space analysts. Their unselfish assistance to me is most appreciated.

Regular news and features about space can be found in the pages of *Flight International* and in *Spaceflight* and *Space World*, publications of the British Interplanetary Society (BIS) and the US National Space Institute (NSI) respectively, both of which the reader is urged to join.

The BIS is at 27/29 South Lambeth Road, London SW8 1SZ and the NSI at West Wing Suite 203, 600 Maryland Avenue SW, Washington DC, USA.

Finally, I would like to thank my wife and fellow worker Sue who retypes and presents my final manuscripts in such great style and without whose support and encouragement I could never have achieved the ambition I had at the time of Gemini 3, to be writing about space flight as an occupation.

Tim Furniss
Epsom, Surrey
25 March 1985

1

Manned Space Flight Diary

Flight 1. (USSR) Vostok 1
12 April 1961 1 h 48 min Yuri Gagarin

After many days of rumour about the impending launch of a Russian into space, the first manned space flight began on 12 April 1961. The spaceman *was* a Russian, Major Yuri Gagarin, and not an American, who was also preparing for a flight, and this proved to be an important factor in dictating the future direction of man in space for many years.

Gagarin had no control over the spacecraft during his flight and spent most of the time sightseeing and reporting how comfortable he felt in weightlessness, a condition that doctors feared would be as dangerous as the high acceleration forces during launch. Gagarin came through it all unscathed and the Vostok 1 capsule landed in a cowfield, while the cosmonaut touched down separately by parachute, having ejected at a height of 22 000 ft, a fact not admitted by the Russians until 1978.

Yuri Gagarin heads for the launch pad with his back-up, Gherman Titov, behind him. Fellow cosmonauts stand watching. It is possible to identify two of them – Bykovsky, third from left, and Nikolyev on the right. (*Novosti*)

Flight 2. (USA 1) MR3 - *Freedom 7*
5 May 1961 15 min 28 s Alan Shepard

Cdr Alan Shepard's flight in a Mercury capsule was not an orbital one but a planned ballistic lob to a height of 116 miles, ending with a splashdown in the Atlantic Ocean. After a 4 h 14 min wait inside the tiny capsule, during which he was compelled to answer the call of nature in his spacesuit, Shepard soared into space, enduring acceleration forces of 11 g. He was weightless for nearly 5 minutes during which time he orientated the craft using gas thrusters, the first spaceman to do so. Shepard only saw the view out of his periscope – and never out of his porthole. The retro rockets which were not needed on this ballistic flight, were test fired all the same and Shepard came down 297 miles from Cape Canaveral to be recovered by a helicopter from the USS *Lake Champlain*.

Flight 3. (USA 2) MR4 - *Liberty Bell 7*
21 July 1961 15 min 37 s Gus Grissom

This flight was planned as a repeat performance of Shepard's which in fact it was – until just after splashdown, when the hatch of the Mercury craft blew open and seawater rushed in. As the capsule sank, Grissom was hauled to safety by a helicopter. He had nearly drowned. The astronaut, who admitted that he was a 'little scared' at lift-off – his pulse peaked at 171 – enjoyed the view out of his window; so much so that he was distracted from his work during the brief period of weightless flight. *Liberty Bell 7* need not have sunk, despite shipping 2000 lb of water. A warning light in the helicopter cockpit was false and the misinformed pilot ditched the spacecraft. Ironically, *Liberty Bell* was the lightest manned spacecraft, weighing 2836 lb in space.

Gus Grissom admitted he felt a 'little scared' at blast off. (*NASA*)

Flight 4. (USSR 2) Vostok 2
6 August 1961 1 day 1 h 18 min
Gherman Titov

Major Gherman Titov was the youngest person in space, at twenty-five, and he still is. He was also the first man to spend a day in space, to sleep and to be sick – the last condition due to the effect of weightlessness on his inner-ear orientation mechanism, a condition subsequently discovered to be fairly common among many spacemen for some period during their flights. Another problem Titov encountered was his floating outstretched arms while he slept, which could have touched some switches on the sparsely populated display panel. He solved this by tucking his arms under his harness. The cosmonaut made a customary ejection and parachute landing.

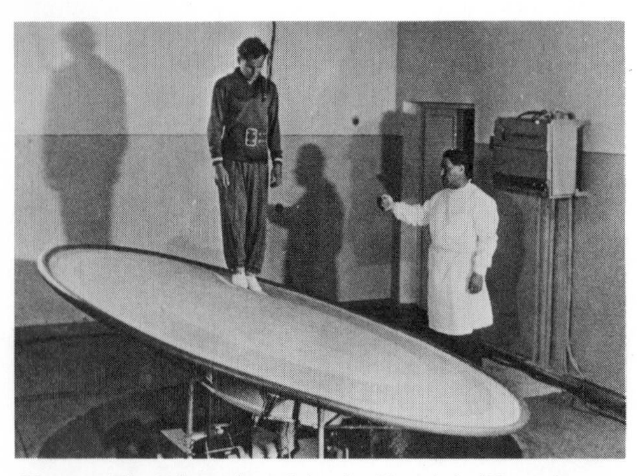

Gherman Titov shown in training for Vostok 2. (*Novosti*)

Flight 5. (USA 3) MA6 - *Friendship 7*
20 February 1962 4 h 55 min 23 s
John Glenn

The first American to orbit the Earth, John Glenn became a national hero, flying three times around the world in a blaze of publicity. This was due in no small part to the man's personality and boy-next-door image and to the fact that the USA was catching up with Russia. The flight was a success, although there was a scary moment when a signal indicated that Glenn's heat shield was loose and might not survive the re-entry. But this proved not to be the case and *Friendship 7* came down in the Atlantic and was hauled aboard the USS *Noa*, with Glenn still inside.

John Glenn walking to the van for the drive to the launch pad on 20 February 1962. (*NASA*)

Flight 6. (USA 4) MA7 - *Aurora 7*
24 May 1962 4 h 56 min 5 s
Scott Carpenter

A fogbound lift-off heralded the beginning of a mission of mixed fortune for Navy Cdr Scott Carpenter. His mainly scientific tasks during the three-orbit flight were well carried out but, unfortunately, later in the mission he became careless and distracted, wasting precious manoeuvring fuel needed to orientate the craft for the re-entry. The retro fire was late and the re-entry was of much concern to the astronaut who landed 125 miles off target and was lost for 36 minutes until he was spotted sitting in his life raft.

Flight 7. (USSR 3) Vostok 3
11 August 1962 3 days 22 h 22 min
Andrian Nikolyev

The first bachelor spaceman, Andrian Nikolyev was sent into space on what was announced to be an extended flight. This would have been achievement enough but little did the public know that the following day Vostok 4 would be in space as well. The first television pictures broadcast from space showed Nikolyev in a cabin that was spacious compared with the US Mercury capsules. During the flight, the cosmonaut also enjoyed the first food not to come out of toothpaste-like tubes; bite-size chunks of cutlets and pies.

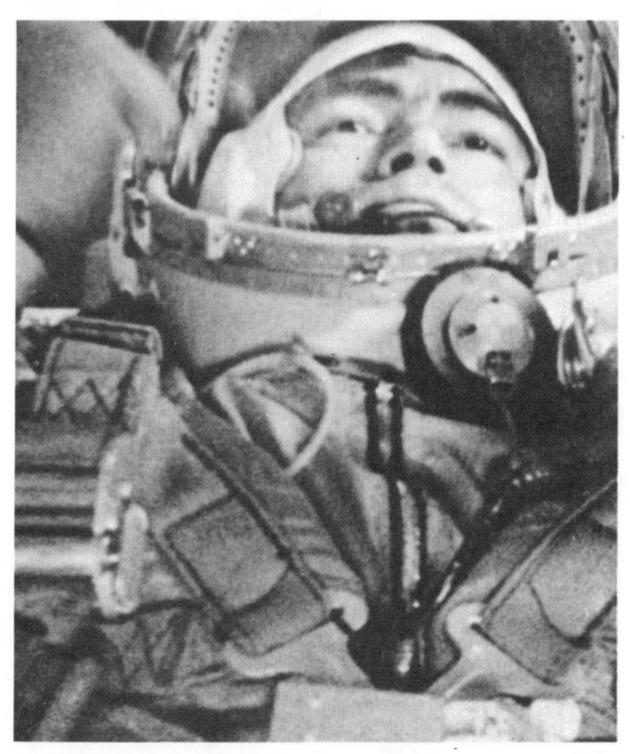

An in-flight photo of Andrian Nikolyev inside Vostok 3. (*Novosti*)

Flight 8. (USSR 4) Vostok 4
12 August 1962 2 days 22 h 57 min
Pavel Popovich

The claim by the Soviet Union that by launching Vostok 4 to within four miles of Vostok 3 during a very brief encounter they had achieved the first space rendezvous was very misleading. So much so that the Western press went overboard, claiming that a Russian would be on the Moon within a few years. The fact is that the Vostoks had no capability to change orbit to effect rendezvous. But the dual flight of the cosmonauts, who were dubbed Nik and Pop, gained a lot of prestige for Russia, still leading the space stakes.

Flight 9. (USA 5) MA8 - *Sigma 7*
3 October 1962 9 h 13 min 11 s
Wally Schirra

The fifth manned Mercury flight was given the modest objective of flying 2¾ extra orbits and drifting as much as possible with little use of the manoeuvring thrusters. Not surprisingly, the flight was described by the astronaut Wally Schirra as 'routine and textbook'. He was able to accomplish everything he had wanted to do, he said. The astronaut splashed down in the Pacific Ocean, notching-up a first in the space programme, which had almost clocked-up another – the first launch abort. Because Schirra's Atlas launcher developed an alarming roll rate, at one point the range safety officer's finger was on the destruct button.

Recovery teams head towards Wally Schirra inside his '*Sigma 7*' Mercury capsule after splashdown in the Pacific. (*NASA*)

Flight 10. (USA 6) MA9 - *Faith 7*
15 May 1963 1 day 10 h 19 min 49 s
Gordon Cooper

This was the final flight of the Mercury programme and the first to last over a day. It was flown by the youngest US astronaut at the time, Gordon Cooper, who performed admirably when he had to take manual control for re-entry due to a malfunction in the craft's automatic control system. Cooper also demonstrated man's ability to see comparatively small objects on the Earth, from 160 miles up. This was due to the clearing effect of the atmosphere to an observer outside it. Cooper could see small log cabins in the Himalayas and the wakes of ships at sea. This was of particular interest to the military services.

Flight 11. (USSR 5) Vostok 5
14 June 1963 4 days 23 h 6 min
Valeri Bykovsky

Valeri Bykovsky flew the longest solo space flight in history but he was totally overshadowed by the flight of a companion spacecraft, Vostok 6. The cosmonaut had been the first to try out many training devices, such as the centrifuge, and was back-up to Nikolyev on Vostok 3. Even before his flight, rumours were emanating from Moscow about the launch of a rather special cosmonaut and while Bykovsky sailed in orbit the world waited.

Valeri Bykovsky, still today the solo space flight record holder. (*Novosti*)

Flight 12. (USSR 6) Vostok 6
16 June 1963 2 days 22 h 50 min
Valentina Tereshkova

Cotton-mill worker and amateur parachutist, Valentina Tereshkova was the latest pawn in Premier Khrushchev's game of space chess with the Americans. The flight of the first lady in space captured the headlines around the world and retained space prestige well and truly with the Soviet Union. No woman followed Tereshkova into space for almost 20 years. Vostok 6 came to within three miles of Vostok 5 in another 'coincidental rendezvous' and Tereshkova was far from happy during her flight. To squeeze as much prestige as possible out of the adventure, Premier Khrushchev arranged the marriage of Valentina to Cosmonaut Nikolyev and although they divorced later the couple did have the first 'space child', Yelena. Tereshkova is still the youngest woman to have travelled in space.

Flight 13. (USSR 7) Voskhod 1
12 October 1964 1 day 0 h 17 min 3 s
Vladimir Komarov, Boris Yegerov, Konstantin Feoktistov

The first three-man spaceship, which captured the headlines so boldly in 1964, was in fact a stripped-down one-man Vostok, with no ejection seats and the crew did not wear spacesuits. The most perilous mission probably ever undertaken, it lasted only a day and the three men – Komarov (the commander), Yegerov (a doctor) and Feoktistov (the scientist whose job it was to convert the Vostok into Voskhod) – could not have been able to do much in the extremely cramped conditions despite being weightless. The craft soft-landed with the crew inside.

Flight 14. (USSR 8) Voskhod 2
18 March 1965 1 day 2 h 2 min 17 s
Pavel Belyayev, Alexei Leonov

On 23 March 1965 the Americans were due to launch the first two-man spacecraft, Gemini, and on a later flight in the programme a man was to walk in space. So five days earlier, the Soviet Union accomplished both achievements during what was to prove their final manned space spectacular for over two years. Alexei Leonov's walk in space was televised and photos of him cavorting in orbit were splashed over the world's front pages. Voskhod 2 made an emergency landing in a forest and the crew had to fight off wolves before being rescued.

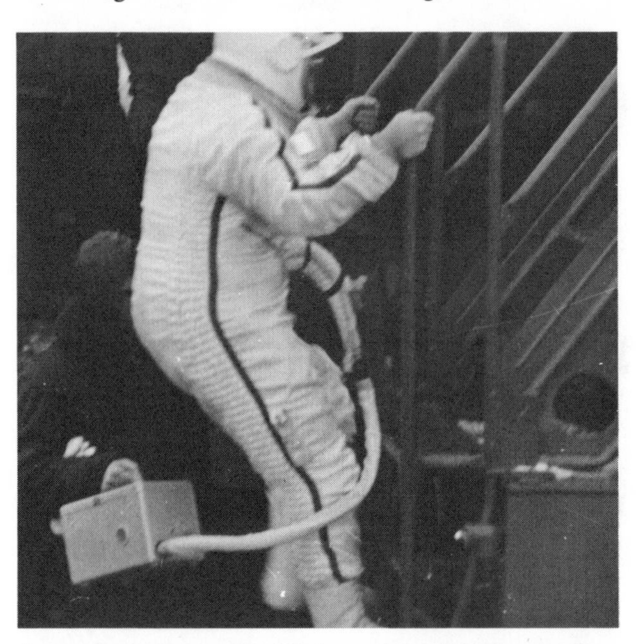

A rare photo showing Pavel Belyayev being helped up the first rungs of a ladder leading to the steps of the launch gantry-tower before the blast-off of Voskhod 2. (*Novosti*)

Flight 15. (USA 7) Gemini 3
23 March 1965 4 h 52 min 51 s
Virgil Grissom, John Young

Gemini 3 became the first craft to make changes to its orbit and the first to use an on-board computer. The test flight was in preparation for ambitious rendezvous and docking attempts to pave the way to the Moon. The spacecraft landed 52 miles off-target and during the long wait in the Atlantic Ocean, Grissom was badly seasick.

Flight 16. (USA 8) Gemini 4
3 June 1965 4 days 1 h 56 min 12 s
James McDivitt, Edward White

After four years behind the Soviet Union, the USA drew level with the spectacular flight of Gemini 4 which featured the first US spacewalk (by Edward White) and the first live television coverage of a launch for viewers in Europe, thanks to the Early Bird comsat. White who provided the subject of some of the best photography to come out of the space programme, walked in space for 22 minutes at the end of a 25 ft tether and for a time used a handheld manoeuvring unit to orientate himself in space.

Flight 17. (USA 9) Gemini 5
21 August 1965 7 days 22 h 55 min 14 s
Gordon Cooper, Charles Conrad

This was the longest manned space flight at the time and an important milestone, as its duration matched the mission time of a flight to the Moon. But after two orbits Gemini 5 almost came home because of a malfunction in fuel cells, being carried on a manned spacecraft for the first time. The fuel cells were designed to provide electrical power. Much of the time the astronauts were reduced to drifting in order to save as much power as possible.

Flight 18. (USA 10) Gemini 7
4 December 1965 13 days 18 h 35 min 1 s
Frank Borman, James Lovell

Because Gemini 6's target rocket for the first rendezvous and docking mission exploded in October, Gemini 7 was launched to act as its manned target for a rendezvous during a long duration study of zero g. Astronauts Borman and Lovell spent two weeks in a cabin no larger than the front seat of a small car but emerged relatively undamaged by weightlessness and re-adaptation to gravity.

After a fortnight in space, Frank Borman, right, and James Lovell acknowledge the welcome from sailors on board the recovery ship. (*NASA*)

Flight 19. (USA 11) Gemini 6
15 December 1965 1 day 1 h 51 min 54 s
Wally Schirra, Thomas Stafford

On 12 December the Titan rocket, with Gemini 6 and its astronauts on top, ignited and within a split second before lift-off was shut down during the first launch pad abort in history. The astronauts were praised for not ejecting. The mission was saved and three days later proved to be a phenomenal success with Gemini 6 and 7 pirouetting together in space during the first space rendezvous. Schirra and Stafford separated from Gemini 7 after 5 h 18 min.

Flight 20. (USA 12) Gemini 8
16 March 1966 10 h 41 min 26 s
Neil Armstrong, David Scott

Neil Armstrong completed the first docking with an Agena target rocket 6½ hours after his launch with David Scott. Soon after however, the mission was heading for disaster as a short-circuited thruster on Gemini fired continuously, sending the spacecraft cartwheeling through space at a rate of 360° a second. Armstrong used so much fuel skilfully bringing Gemini under control, that an emergency landing had to be made.

Flight 21. (USA 13) Gemini 9
3 June 1966 3 days 0 h 20 min 50 s
Thomas Stafford, Eugene Cernan

Gemini 9 was a flight dogged by ill-luck. Firstly its prime crew was killed in an air crash. Secondly, the Agena 9 target rocked exploded. Thirdly, a secondary target called the ATDA failed to shed its payload shrouds protecting the docking collar, so a docking could not take place. Fourthly, Gemini was stranded on the pad on 1 June within two minutes of blast off. Finally, after a successful rendezvous Eugene Cernan's EVA, during which he was to have flown a manoeuvring unit, had to be cut short. Gemini 9 did however make a bulls-eye splashdown within a mile of the recovery ship. Cernan is the youngest American male in space.

Flight 22. (USA 14) Gemini 10
18 July 1966 2 days 22 h 46 min 39 s
John Young, Michael Collins

This taciturn pair of astronauts were ordered by mission control to chat more in order to please the PR people. Gemini 10 made a successful docking with Agena 10 and, for the first time in history, used Agena's engine to boost it to a height of 474 miles. Young kept a rendezvous with the Agena 8 target and Collins, during a 39 minutes curtailed EVA, became the first man to make bodily contact with another craft in space when he flew over to it.

Gemini 9's Eugene Cernan, left, and Tom Stafford nonchalantly wait for their collection by an aircraft carrier recovery ship. (*NASA*)

Agena 10 seen from Mike Collins's window prior to docking. (*NASA*)

Flight 23. (USA 15) Gemini 11
12 September 1966 2 days 23 h 17 min 8 s
Charles Conrad, Richard Gordon

Gemini 11 docked with its target rocket during the first orbit and using the Agena's engine was boosted to a record height of 850 miles over Australia. From a height of 500 miles, the astronauts took a classic photo of India and Sri Lanka. During an EVA, yet again curtailed by problems, Gordon attached a tether to Agena which was later used in an experiment to create artificial gravity in space. Gemini 11 made a totally automatic retrofire and re-entry, splashing down within 1½ miles of the recovery ship.

Gemini 11's view of Australia from a record orbital height of 850 miles showing an area from Perth to Port Darwin. (*NASA*)

Flight 24. (USA 16) Gemini 12
11 November 1966 3 days 22 h 34 min 31 s
James Lovell, Edwin Aldrin

At last, Gemini experienced a fully successful EVA when Aldrin spent 2 h 8 min outside the spacecraft in another important milestone on the road to the Moon. Preparations for Apollo were concluded by Gemini 12 which made another docking with a target rocket, although not a re-boost. The Gemini programme clocked up 80 man-days in space and everything was ready for Apollo.

APOLLO 1
27 January 1967
Virgil Grissom, Edward White, Roger Chaffee

Apollo was scheduled for a thorough shakedown in Earth orbit on 21 February 1967. On 27 January the three astronauts were conducting a launch countdown demonstration test on Pad 34, when a short circuit in an electrical wire caused a spark which in a pure 100 per cent oxygen atmosphere turned into an inferno in which the astronauts perished. The Apollo programme was delayed over a year but, without the disaster and the faults which the investigation into it unearthed, it is unlikely that man would have reached the Moon when he did.

Killed in the horrific Apollo 1 spacecraft fire were Gus Grissom left, Roger Chaffee and Ed White. (*NASA*)

Flight 25. (USSR 9) Soyuz 1
27 April 1967 1 day 2 h 48 min
Vladimir Komarov

The mission plan for this flight involved the lone Komarov acting as a target for a crew of three on Soyuz 2. But before the second mission could begin, Soyuz 1 was in trouble. Several malfunctions caused plans for an emergency return to Earth but Komarov, the first Russian to enter space twice, did not succeed in firing the retro rockets until the 18th orbit. The Russians say that a parachute failure caused the Soyuz capsule to plummet to Earth but Komarov might have already been killed during the high temperature re-entry which must have had some effect on the tumbling craft. Like Apollo, Soyuz was delayed a year.

Flight 26. (USA 17) Apollo 7
11 October 1968 10 days 20 h 9 min 3 s
Wally Schirra, Donn Eisele,
Walt Cunningham

A flight that was critical to America's chances of reaching the Moon by the end of the sixties, Apollo 7 was designed to test out fully the command and service modules in Earth orbit. The flight was termed as 101 per cent successful by NASA although the astronauts' behaviour verged at times on the curt and surly. The Service Propulsion System engine was fired eight times and once for 66 seconds in a simulation of lunar orbit insertion and de-orbit burns. At splashdown, the Apollo capsule turned upside down but was soon righted by bouyancy bags.

Flight 27. (USSR 10) Soyuz 3
26 October 1968 3 days 22 h 51 min
Georgi Beregovoi

Cosmonaut Beregovoi, then the oldest man in space at forty-seven, was to perform a rendezvous and docking with the unmanned Soyuz 2 but failed. His launch was the first to be shown on television and during his flight, which was the second longest in the Russian programme, Beregovoi televised many 'shows' from space. At the same time the Soviets were sending unmanned Soyuz craft, called Zond, on lunar looping flights in a bid to send one cosmonaut to the Moon before the Americans.

Flight 28. (USA 18) Apollo 8
21 December 1968 6 days 3 h 0 min 42 s
Frank Borman, James Lovell,
William Anders

Apollo 8 was originally planned to be a high altitude test in Earth orbit but the obvious impending Soviet attempt to send one cosmonaut to the Moon caused the mission to be turned into a lunar orbit one. A spectacular flight, probably the most remembered and certainly the most publicized, featured the readings from the Book of Genesis on Christmas Day by the three astronauts orbiting the Moon. It was the first manned mission atop the mighty Saturn 5.

Flight 29. (USSR 11) Soyuz 4
14 January 1969 2 days 23 h 21 min
Vladimir Shatalov

Soyuz 4 was the target craft for the ambitious rendezvous and docking attempt that was to have taken place two years earlier. Vladimir Shatalov became the first spaceman to play host to visitors, two crewmen from Soyuz 5, who both landed in Soyuz 4.

Flight 30. (USSR 12) Soyuz 5
15 January 1969 3 days 0 h 54 min
Boris Volynov, Alexei Yeliseyev,
Yevgeni Khrunov

Soyuz 5 achieved a perfect docking with Soyuz 4 on 16 January after 17 orbits of the Earth and soon afterwards cosmonauts Yeliseyev and Khrunov, performing the first dual EVA, transferred to Soyuz 4, delivering post and ceremonial documents to Shatalov. Volynov flew solo and landed with a flight time of 3 days 0 h 54 min while his original passengers, the first to land in a different craft from the one in which they were launched, clocked up a flight time of just under two days.

Flight 31. (USA 19) Apollo 9
3 March 1969 10 days 1 h 0 min 54 s
James McDivitt, David Scott,
Russell Schweickart

The Apollo lunar module was tested by astronauts for the first time during an Earth orbital flight. McDivitt and Schweickart made a simulated landing on the Moon and take off, later rendezvousing and docking safely with the Command Module. Earlier, although recovering from a bout of space sickness, Schweickart performed an EVA from the lunar module to test out the Apollo space suit and its unique portable life-support system backpack.

Flight 32. (USA 20) Apollo 10
18 May 1969 8 days 0 h 3 min 23 s
Thomas Stafford, John Young,
Eugene Cernan

This flight was originally slated to be the first landing on the Moon and Tom Stafford would have been the first man to walk on the lunar surface. But the Lunar Module was deemed to be unsuitable for a landing and a further test flight was thought necessary. So, Stafford and Cernan flew a complete

Apollo 11

Flight 33. (USA 21) Apollo 11
16 July 1969 8 days 3 h 18 min 35 s
Neil Armstrong, Michael Collins,
Edwin Aldrin

President Kennedy set the USA on a course to the Moon on 25 May 1961. His goal was achieved within 5½ months of the deadline he set. Millions of people all over the world watched on television as the ghostly figures of Armstrong and Aldrin walked on the surface of the Moon. Ironically, the two current affairs events that most people can relate to exactly what they were doing at the time were Kennedy's assassination and the Moon walk. The landing achieved, amid a sense of anti-climax, Apollo began to lose favour, although ten more men went to the Moon.

Edwin Aldrin descends the LM ladder towards the lunar surface. (*NASA*)

The flight patch
of the historic mission
of Apollo 11. (*NASA*)

landing mission simulation coming to within nine miles of the surface, while John Young became the first man to fly solo in lunar orbit. A safe return to Earth gave the green signal to Apollo 11. Apollo 10 also achieved the highest speed attained by man: 24 790 mph.

Flight 34. (USSR 13) Soyuz 6
11 October 1969 4 days 22 h 42 min
Georgi Shonin, Valeri Kubasov

The first of three spacecraft in a strange troika mission which didn't seem to achieve anything other than mark the first occasion in which eight men were in space at once, Soyuz 6 featured a unique welding experiment called Vulkan in the craft's orbital module. Shonin moved to within a mile of Soyuz 7 and 8 to observe what is assumed to have been an attempted docking à la Soyuz 4/5.

Flight 35. (USSR 14) Soyuz 7
12 October 1969 4 days 22 h 41 min
Anatoli Filipchenko, Vladislav Volkov, Viktor Gorbatko

It is assumed that had Soyuz 7 been able to dock with Soyuz 8 – the nearest they actually got to each other was 1600 ft – an internal crew transfer through a docking tunnel would have taken place. The three craft, 6, 7 and 8, remained close to each other for about a day before moving away for separate experiment flights.

Inside Soyuz 7 are, left to right, Vladislav Volkov, Anatoli Filipchenko and Viktor Gorbatko. Note, the cosmonauts are not wearing spacesuits. (*Novosti*)

Flight 36. (USSR 15) Soyuz 8
13 October 1969 4 days 22 h 51 min
Vladimir Shatalov, Alexei Yeliseyev

Soyuz 4/5 group flight cosmonauts teamed up for another flight as lead crew of the 6, 7, 8 troika flight. The mission of the three Soyuz craft possibly resulted from the cancellation of the manned Zond mission to the Moon, leaving spare craft available or perhaps from failures of the first Salyut space stations, also leaving ferry craft available.

Flight 37. (USA 22) Apollo 12
14 November 1969 10 days 4 h 36 min 25 s
Charles Conrad, Richard Gordon, Alan Bean

Launched in a thunderstorm, Apollo 12 was struck by lightning during its ascent but survived to make a pinpoint landing on the Moon, 553 feet from the unmanned Surveyor 3, which had made contact in 1967. During two Moon walks, unfortunately latterly not televised because the camera was pointed at the Sun early on and blacked out, Conrad and Bean sang and whistled their way about, collecting rocks and laying down an array of surface experiments.

Alan Bean at work on the Moon during the mission of Apollo 12. (*NASA*)

Flight 38. (USA 23) Apollo 13
11 April 1970 5 days 22 h 54 min 41 s
James Lovell, Jack Swigert, Fred Haise

A routine, relatively unpublicized, engineering 'milk run' to the Moon was transformed on 13 April at T + 55 h 55 min 20 s into the most dramatic space mission in history. An explosion of an oxygen tank in Apollo's service module crippled this and the Command Module's systems. The astronauts would have died had they been returning from the Moon and not going towards it but were able to use the propulsion and other systems of the Lunar Module Aquarius to limp home. Millions of people all over the world breathed a sigh of relief when the crippled command module just made it home. The astronauts' fly-by of the Moon made them the deepest voyagers into space 248 655 miles from Earth.

The world breathed a sign of relief when the Apollo 13 capsule splashed down. (*NASA*)

Flight 39. (USSR 16) Soyuz 9

1 June 1970   17 days 16 h 58 min 50 s
Andrian Nikolyev, Vitali Sevastyanov

Delays in the establishment of the Salyut space station led to the solo flight of another spare Soyuz. The first Soviet extended mission, it was highly significant in that it revealed the problems that man in space will always experience, not necessarily related to zero g itself but to readaptation to gravity on his return to Earth after a long period in space. The cosmonauts suffered badly for almost two weeks after landing and it became obvious that steps had to be taken to reduce this readaptation stress while the cosmonaut was still in orbit by, among other things, strict, rigorous exercise.

Vitali Sevastyanov, left, and Andrian Nikolyev inside Soyuz 9 during their flight. (*Novosti*)

Flight 40. (USA 24) Apollo 14

31 January 1971   9 days 0 h 1 min 57 s
Alan Shepard, Stuart Roosa,
Edgar Mitchell

The most inexperienced Apollo crew to take to the skies, commanded by 47-year-old Alan Shepard with just 15 minutes' space experience under his belt, performed admirably during a highly successful scientific mission to Fra Mauro. Although Apollo Lunar Surface Experiments were deployed on the surface and 98 lb of Moon rock were collected, most people remember Shepard's game of lunar golf as the highlight of the mission. Shepard is the oldest man on the Moon.

Alan Shepard, commander of Apollo 14, takes a last look at his Saturn 5 launcher before blast-off to the Moon. (*NASA*)

Flight 41. (USSR 17) Soyuz 10

23 April 1971   1 day 23 h 45 min
Vladimir Shatalov, Alexei Yeliseyev,
Nikolai Ruchavishnikov

The Soviet Union at last managed to get a Salyut space station safely into orbit on 19 April 1971 and four days later sent a three-man crew on Soyuz 10 up to dock with it. A long-duration mission of perhaps three weeks was on the agenda. Shatalov docked Soyuz 10 after 90 minutes' effort, during which it is thought the transfer tunnel must have been damaged, because after 5½ hours together, Soyuz and Salyut were parted and the rueful crew came home.

FLIGHT 42. (USSR 18) Soyuz 11

6 June 1971   23 days 18 h 22 min
Georgi Dobrovolsky, Vladislav Volkov,
Viktor Patsayev

Soyuz 11 succeeded in docking with Salyut 1 and its crew actually got inside this time. There they remained for a highly successful three week course in scientific and basic science experiments. The mission complete, Soyuz 11 backed off, fired retro rockets and the orbital compartment containing the docking mechanism was ejected. The explosive separation sequence shook open an exhaust valve in the cabin and within 45 seconds the cosmonauts were dead. Soyuz made a perfect landing. If the cosmonauts had been wearing space suits they most probably would have survived.

Left to right, Viktor Patsayev, Vladislav Volkov and Georgi Dobrovolsky died because they were not wearing spacesuits when Soyuz 11 depressurized. (*Novosti*)

Flight 43. (USA 25) Apollo 15

26 July 1971 12 days 7 h 11 min 53 s
David Scott, Alfred Worden, James Irwin

Dave Scott made a steep 26° descent over the Appenine Mountains of the Moon to make a perfect landing on Hadley Plain close to the famous rille, or valley. Here, he and Irwin completed three lunar excursions and also drove the first lunar roving vehicle which was deployed from the side of their lunar module Falcon. During the return flight to Earth, Worden made a spacewalk to retrieve science packages from the side of the service module. One of the three parachutes on the command module failed just before splashdown but the astronauts were unhurt. With the SIVB stage attached, prior to the trans lunar insertion burn, Apollo 15 was the heaviest vehicle in Earth orbit at 309 828 lb.

An eve-of-launch photo of Apollo 15. (*Author*)

Flight 44. (USA 26) Apollo 16

16 April 1972 11 days 1 h 51 min 5 s
John Young, Thomas Mattingly, Charles Duke

Failure of the back-up manual control system in the Apollo service module propulsion system nearly aborted this flight just before lunar landing. A few hours late, Young and Duke landed safely on the highest point on the Moon so far reached by man – at Descartes. The astronauts gambolled around the surface in a highly enthusiastic manner, providing entertainment for the now dwindling television audiences. After three lunar walks and a 'grand prix' test of the roving vehicle, the astronauts left the Moon to join Mattingly who flew the longest US solo flight and who later made an EVA during the return journey. Duke is the youngest man on the Moon at thirty-six.

Flight 45. (NASA 27) Apollo 17

7 December 1972 12 days 13 h 51 min 59 s
Eugene Cernan, Ronald Evans, Harrison Schmitt

This was the final flight to the Moon because three Apollo flights had been cancelled as a result of dwindling public and political support. As such, it naturally flew the only astronaut-geologist in the NASA space corps, Harrison 'Jack' Schmitt, who had to replace a test pilot from the crew to do so. The launch was spectacular, taking place at night, the Saturn 5 turning night into day for miles around. At the Taurus Littrow landing site Cernan and Schmitt conducted three walks and drove the lunar rover for a total of 21 miles. Apollo 17 was probably the last flight to the Moon this century. Cernan's third lunar walk was the longest EVA on record at 7 h 37 min.

The last man on the Moon, Eugene Cernan, bids farewell to the pad team before entering Apollo 17. (*NASA*)

Flight 46. (USA 28) Skylab 2
25 May 1973 28 days 0 h 49 min 49 s
Charles Conrad, Joseph Kerwin,
Paul Weitz

Skylab, like Salyut, was a first-generation space station and it was built from leftover components from Apollo, to enable astronauts to remain in space for a maximum of two months to assess their ability to work and live in space. Skylab was severely damaged during launch but the three astronauts of Skylab 2 – an Apollo command and service module – saved the entire programme during their record-breaking mission. Conrad and Kerwin, a medical doctor, performed a hazardous EVA during which they prized out a jammed solar panel.

Flight 47. (USA 29) Skylab 3
28 July 1973 59 days 11 h 9 min 4 s
Alan Bean, Owen Garriott, Jack Lousma

This mission exceeded the programme's endurance target by three days during an action-packed flight full of scientific and medical experiments. The Apollo command and service module suffered a failure while attached to Skylab and a rescue flight was being prepared before the crisis was overcome. The flight also saw the first serious cases of space sickness to hit US astronauts but because of their exercise routine while in orbit their readaptation to gravity was not difficult.

Flight 48. (USSR 19) Soyuz 12
27 September 1973 1 day 23 h 16 min
Vasili Lazarev, Oleg Makarov

This was the maiden flight of a new type of Salyut ferry vehicle; a Soyuz without the solar panels required for extended independent flight but with enough power to perform a rendezvous and docking with the space station within two days. Other refinements included not three crew but two, each wearing space suits. For the first time, the Soviets announced how long the crew would stay in space – two days – to avert rumours of a failure in the West.

Flight 49. (USA 30) Skylab 4
16 November 1973 84 days 1 h 15 min 31 s
Gerald Carr, Edward Gibson,
William Pogue

The longest US space mission ever undertaken, this final flight aboard the Skylab space station was initially marred by space sickness and mutiny by the crew who covered up mistakes. Later, however, things went swimmingly and the crew conducted 56 experiments and 26 science demonstrations; they also studied the Sun for 338 hours. Although this was the last flight by Americans before the space shuttle, apart from the US/USSR mission in 1975, Skylab did briefly enter the limelight again during a highly spectacular re-entry in 1979.

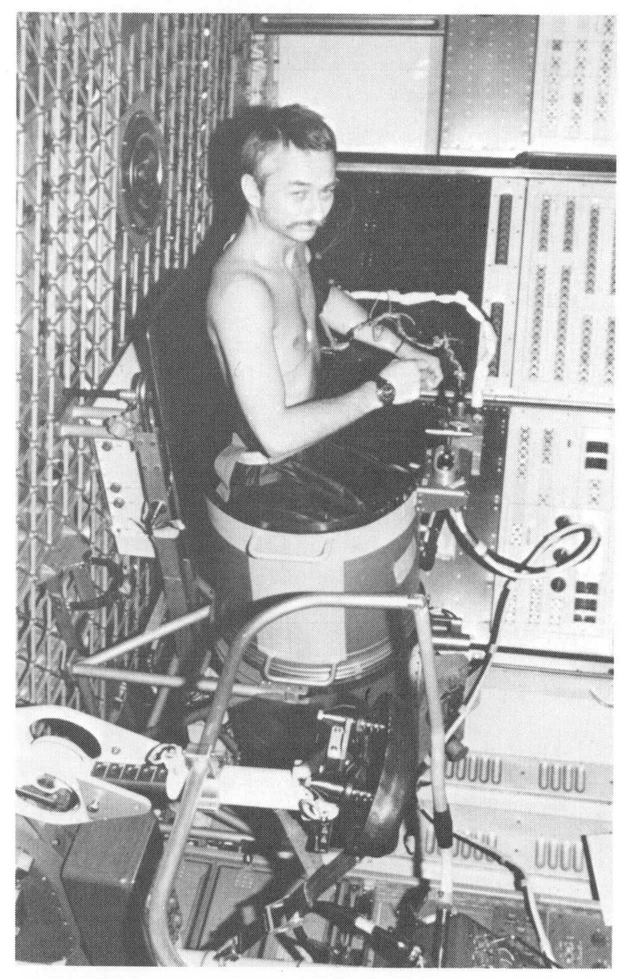

Owen Garriott aboard the Skylab space station wearing a lower body negative g suit to impose gravity on his body in zero g. (*NASA*)

The US long-duration space record holders, left to right, Gerald Carr, Edward Gibson and Bill Pogue. (*NASA*)

Flight 50. (USSR 20) Soyuz 13
18 December 1973 7 days 20 h 55 min
Pyotr Klimuk, Valentin Lebedev

Soyuz 13 was an old-style Soyuz spacecraft which was not designed for ferry flights to the Salyut space station but to fly independently during an eight-day science mission. The mission coincided with a manned occupation of the US Skylab space station and marked the first time that both Americans and Russians were in space together – although they never met. The main payload of Soyuz 13 were the Orion celestial telescope and Oasis protein manufacturing unit. Geoffrey Perry of Kettering Grammar School announced the end of the flight one hour before the Russians did.

Flight 51. (USSR 21) Soyuz 14
3 July 1974 15 days 17 h 30 min
Pavel Popovich, Yuri Artyukhin

Salyut 3 was a military space station equipped with a 33 ft focal length optical telescope. Its military crew – the first official space spies – included Vostok veteran, Pavel Popovich. Although the Soviets announced details of the state of the crew, no information was given as to what they actually did during the 2 week mission. A massive solar storm sent streams of potentially dangerous radiation towards the Earth from 5 to 8 July and at the time the crew prepared to come home early, but the situation was not quite as serious as first thought.

Flight 52. (USSR 22) Soyuz 15
26 August 1974 2 days 0 h 12 min
Gennadi Sarafanov, Lev Demin

This flight of the new-style Soyuz ferry vehicle failed. The craft had not made contact with Salyut 3 within the prescribed 2 days, due to uncontrollable and excessive burns during its final approach, so had to be ordered home. The military crewmen, including grandfather Lev Demin, forty-eight, made an emergency landing at night.

Flight 53. (USSR 23) Soyuz 16
2 December 1974 5 days 22 h 24 min
Anatoli Filipchenko,
Nikolai Ruchavishnikov

In 1972 the Soviet Union and the USA agreed to fly a joint manned space mission involving Soyuz and Apollo spacecraft. A special docking mechanism was developed and fitted to the Soyuz orbital module and Apollo's command module. Soyuz 16 was a test flight to check out this new device and other systems, although no actual docking took place. It was probably flown to give the USA confidence in the Soviet spacecraft after a spate of failures within the Soyuz and Salyut programmes.

Flight 54. (USSR 24) Soyuz 17
11 January 1975 29 days 13 h 20 min
Alexei Gubarev, Georgi Grechko

A civilian scientific research space station, Salyut 4 was launched on 26 December 1974 and, soon after, its first resident crew was aboard for a record Russian flight of over a month. The crewmen began a series of important exercises to impose pressure on their bodies to help ease readaptation to gravity on their return. Although re-entry was uneventful the landing wasn't. The cloud deck was only 800 ft and winds were 44 mph.

Flight 55. (USSR 25) Soyuz 18-1
5 April 1975 21 min 27 s Vasili Lazarev, Oleg Makarov

All seemed to be proceeding well with the launch of this new visiting crew to Salyut 4 until the second stage of the A2 rocket ignited . Unfortunately, the first stage and the second had not separated completely and the vehicle went out of control. Soyuz was ejected 4 seconds later and during the emergency landing from about 90 miles altitude, Lazarev and Makarov endured a high g load of 14 g. The capsule landed on a mountain side and only the parachute line snarling a tree saved the crew.

Flight 56. (USSR 26) Soyuz 18
24 May 1975 62 days 23 h 20 min
Pyotr Klimuk, Vitali Sevastyanov

This record-breaking flight by the Soviets involved the taking of 2000 Earth resources photographs and the discovery of a black hole in the constellation Cygnus. In all, 13 days were spent on geophysical work, 13 on astrophysics, 6 on technical, 10 medical, 2 photography, 2 on atmospheric experiments, 7 on packing and unpacking and 10 days of relaxation. The crew's separation from Salyut 4 and their landing were shown live on television.

At work inside Salyut 4 during the mission of Soyuz 18 are, left, Pyotr Klimuk and Vitali Sevastyanov. (*Novosti*)

Flight 57. (USSR 27) Soyuz 19
15 July 1975 5 days 22 h 3 min 54 s
Alexei Leonov, Valeri Kubasov

Soyuz 19 – representing the Soviets in the extravaganza of détente, ASTP – was launched first at 3.20 pm Moscow time while 10 000 miles away Apollo waited on Pad 39B. The spacecraft was on its 36th orbit when Apollo's docking system merged with Soyuz. 'Soyuz and Apollo are shaking hands now', said the English-speaking Leonov. Many ceremonial events took place and at one time American astronauts were flying in Soyuz. Leonov and Kubasov landed first.

Flight 58. (USA 31) Apollo 18
15 July 1975 9 days 1 h 28 min 24 s
Thomas Stafford, Vance Brand,
Donald 'Deke' Slayton

Fifty-one-year-old Deke Slayton, the only Mercury astronaut who never flew the spacecraft, finally made it after a wait of 16 years with NASA as docking module pilot of the ASTP mission. Apollo 18 did most of the work for the rendezvous and docking and after separation from Soyuz continued in space for a while longer. Disaster almost struck just before splashdown when the crew was gassed by nitrogen tetroxide from the reaction control systems thrusters. Brand was unconscious at splashdown. But no lasting harmful effects befell the astronauts.

The veterans of ASTP. Tom Stafford, left, and Alexei Leonov. (*Novosti*)

Flight 59. (USSR 28) Soyuz 21
6 July 1976 49 days 6 h 24 min
Boris Volynov, Vitali Zholobov

After almost a year's hiatus in manned space flight and with a new military station (Salyut 5) in orbit, all was set for the launching of Soyuz 21 to meet it. During their extended stay in space, Volynov and Zholobov monitored air and sea movements in Siberia, employing the operation Sevier military manoeuvres to assess their ability to spy from space. The flight ended in emergency evacuation when an acrid odour was emitted by Salyut's environmental control system.

Flight 60. (USSR 29) Soyuz 22
15 September 1976 7 days 21 h 52 min
Valeri Bykovsky, Vladimir Aksyonov

This was the back-up craft for ASTP and the docking system was replaced by a multispectral camera, manufactured by Karl Zeiss of East Germany, which carried out detailed Earth resources photography as part of an experiment called Raduga. Two thousand four hundred photos were taken covering 30 specific targets including military ones in East Germany, northern USSR and Norway, where a NATO exercise was being conducted.

Flight 61. (USSR 30) Soyuz 23
14 October 1976 2 days 0 h 6 min
Vyacheslav Zudov,
Valeri Rozhdestvensky

The rendezvous approach electronics failed just as Soyuz 23 was approaching Salyut 5 for the second military flight. The ferry vehicle had to make an emergency landing, performing the first Soviet splashdown into the ice-filled Lake Tengiz in the dead of night, in high winds and in a temperature of about −20 °C. According to the Soviets, a 'certain amount of heroism' was involved in a dramatic rescue of the cosmonauts.

Flight 62. (USSR 31) Soyuz 24
7 February 1977 17 days 17 h 26 min
Viktor Gorbatko, Yuri Glazkov

Although this mission was a military one – the last dedicated military flight to date by the Soviets – the crew did engage in some scientific and technical work during their comparatively short stay. Soldering and casting of metals was carried out along with crystal growth experiments and solar and Earth resources photography.

Alexei Leonov, left, with the crew of Soyuz 24, Viktor Gorbatko, centre, and Yuri Glazkov. In the background is the A2 second stage. (*Novosti*)

Flight 63. (USSR 32) Soyuz 25
9 October 1977 2 days 0 h 46 min
Vladimir Kovalyonok, Valeri Ryumin

A year after the failure of the ferry craft Soyuz 23, a new craft arriving to inaugurate operations on a new space station, Salyut 6, failed to dock correctly and had to be ignominiously brought home for an emergency landing, making rather a farce of all the pre-launch publicity surrounding the 'new era' that was to be opened by Salyut 6. It was the eighth Russian space station mission to fail in 13 attempts.

Flight 64. (USSR 33) Soyuz 26
10 December 1977 96 days 10 h
(landed in Soyuz 27) Yuri Romanenko, Georgi Grechko

Soyuz 26 docked at a secondary docking port at the rear of Salyut 6 to begin a record-breaking flight of over 90 days. During a spacewalk to inspect the primary docking port, thought to have been damaged by Soyuz 25, Grechko, tethered to Salyut, saved Romanenko's life when the eager and untethered cosmonaut leaned out too far, lost his grip and floated out of the space station. The cosmonauts received visitors from Soyuz 27, 28 and a new unmanned tanker craft called Progress. The American endurance record was beaten on this flight.

Yuri Romanenko, left, and Georgi Grechko of Soyuz 26 in front of a Salyut simulator. (*Novosti*)

Flight 65. (USSR 34) Soyuz 27
10 January 1978 5 days 22 h 59 min
(landed in Soyuz 26) Vladimir Dzhanibekov, Oleg Makarov

The first visitors to Salyut 6 were dragged inside by the enthusiastic residents for a short stay on board to carry out medical experiments. Dzhanibekov gave the station's electrical system a check. Soyuz 27 was the first re-supply vehicle and the crew handed over newspapers, letters, books and research equipment. Later they swopped space ships for landing.

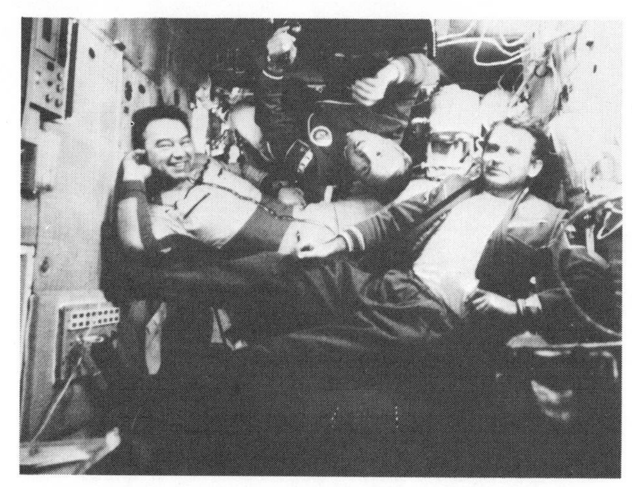

Grechko, left, with Vladimir Dzhanibekov, centre, and Oleg Makarov during their visit to Salyut 6 on Soyuz 27. (*Novosti*)

Flight 66. (USSR 35) Soyuz 28
2 March 1978 7 days 20 h 16 min
Alexei Gubarev, Vladimir Remek

Seventeen years into the space age the first non-American, non-Russian spaceman was launched: Czechoslovakia's Vladimir Remek was the first of many Soviet-bloc cosmonauts to fly Intercosmos missions to Salyut space stations. He and Gubarev were the second visitors to Romanenko and Grechko, now into their third month in space. Many Czech experiments were operated by Remek including a unit to make super pure crystals in zero gravity.

Flight 67. (USSR 36) Soyuz 29
15 June 1978 139 days 14 h 48 min
(landed in Soyuz 31) Vladimir Kovalyonok, Alexander Ivanchenkov

During this crew's 4½ months in Salyut it hosted two Intercosmos visits and received three visits from Progress tankers. Kovalyonok and Ivanchenkov carried out major Earth resources survey work and tests to verify semi-conductor production techniques. Ivanchenkov made a spacewalk to retrieve rubber, polymer and biopolymer samples from the side of Salyut to enable scientists to assess the effects of space exposure on these materials.

Flight 68. (USSR 37) Soyuz 30
27 June 1978 7 days 22 h 4 min
Pyotr Klimuk, Miroslaw Hermaszewski

Polish cosmonaut Hermaszewski operated an experimental unit in Salyut 6 called Serena which manufactured cadmium-tellurium-mercury semi-conductor material, the most sensitive known detector of infra-red radiation and worth £4000 per gram. The mission followed the set routine pattern of Intercosmos visits to Salyut.

Flight 69. (USSR 38) Soyuz 31
26 August 1978 7 days 20 h 49 min
(landed in Soyuz 29) Valeri Bykovsky, Sigmund Jähn

Jähn, the East German cosmonaut, acted as chief photographer in Salyut 6, using the MKF-6M Earth resources multispectral camera, once during a military exercise. This was another impressively routine mission and one which attracted little publicity.

Flight 70. (USSR 39) Soyuz 32
25 February 1979 175 days 0 h 36 min
(landed in Soyuz 34) Vladimir Lyakhov, Valeri Ryumin

Because a planned visit to Salyut by a Bulgarian cosmonaut failed and later visits by a Hungarian and a Cuban were cancelled on the grounds of cautious safety, the long duration crew of Lyakhov and Ryumin spent the entire mission on their own. The most successful Soviet mission flown to that date, Soyuz 32 notched up many firsts, including the unjamming of a radio telescope outside Salyut by Ryumin during a hazardous EVA. Because there were fears that Soyuz 32 had exceeded its safe design life, an unmanned Soyuz 34 was sent up to bring the crew home.

After 175 days in space, Valeri Ryumin, left, and Vladimir Lyakhov have their first unkind taste of gravity after landing from their Soyuz 32 mission, in Soyuz 34. (*Novosti*)

Flight 71. (USSR 40) Soyuz 33
10 April 1979 1 day 23 h 1 min
Nikolai Ruchavishnikov, Georgi Ivanov

The Bulgarian Intercosmos flight failed to dock with Salyut 6. Soyuz 33's primary engine failed and because of the gravity of the situation the back-up engine was fired for 213 seconds to initiate an immediate one phase re-entry at high g, at 15 +, possibly the highest endured by spacemen in flight. The capsule glowed red as it descended to Earth against the darkening sky. The commander was the first non-pilot to lead a space mission.

Space Shuttle Columbia

Flight 80. (USA 32) STS1
12 April 1981
2 days 6 h 20 min 52 s
John Young, Robert Crippen

The long-awaited maiden flight of the space shuttle *Columbia*, delayed for over two years, was a remarkable success. The sight of a spaceship returning to Earth like an airliner captured the imagination of the world and opened a new era of space commercialization for America. Tiles that fell from *Columbia* during launch, a loss witnessed by television viewers when the craft had achieved orbit, marred the flight, not because it was a critical situation but because the popular press, eager to knock the shuttle which had many critics at this time, made such a fuss of it.

The long awaited maiden launch of the revolutionary, reusable Space Shuttle. (*NASA*)

Astronauts in T-38 jet fighter monitor lauch of STS1. (*NASA*)

Flight 72. (USSR 41) Soyuz 35
9 April 1980 184 days 20 h 12 min
(landed in Soyuz 37) Leonid Popov,
Valeri Ryumin

Ryumin, fresh from 175 days in space, was not originally slated for Soyuz 35 but when its flight engineer was injured he stepped in to make yet another arduous long duration flight and to become a world-record holder. He and Popov played hosts to Intercosmos crews from Hungary, Vietnam and Cuba, a crew of a new craft, Soyuz T2 and two Progress tanker vessels. The cosmonauts unusually returned nearly 10 lb heavier thanks to a 'well ordered life style'.

Flight 73. (USSR 42) Soyuz 36
26 May 1980 7 days 20 h 46 min
(landed in Soyuz 35) Valeri Kubasov,
Bertalan Farkas

Hungarian cosmonaut Farkas arrived at Salyut bearing a meal of goulash, pâté de foie gras, fried pork and jellied tongue, plus an array of experiments, one of which manufactured gallium arsenide crystals with chromium. Another experiment was used to study the possibility of manufacturing the cancer-fighting drug Interferon in space.

Flight 74. (USSR 43) Soyuz T2
5 June 1980 3 days 22 h 19 min
Yuri Malyshev, Vladimir Aksyonov

This new-style Soyuz ferry was uprated to perform automatic dockings with Salyut but ironically these systems failed and Malyshev performed a manual docking. The crew wore new-style spacesuits and on the way home jettisoned the orbital module before retro fire to save fuel and to demonstrate that the module could be left attached to Salyut to enlarge its capacity.

Vladimir Aksyonov tries out the new Soyuz T spacesuit prior to his mission on Soyuz T2. (*Novosti*)

Flight 75. (USSR 44) Soyuz 37
23 July 1980 7 days 20 h 42 min
(landed in Soyuz 36) Viktor Gorbatko,
Pham Tuan

Amid more than the usual welter of propaganda normally associated with the Intercosmos flights, Vietnamese Pham Tuan was hailed as the only pilot to have shot down an American B52 during the Vietnam War. Tuan suffered early in the flight from slight nausea, loss of appetite and a headache. As usual, a lot of MKF-6M Earth resources photography of the Intercosmos pilot's home country was carried out.

One of the many Intercosmos missions ends with a smiling Pham Tuan from Vietnam and Viktor Gorbatko speaking to the press. (*Novosti*)

Flight 76. (USSR 45) Soyuz 38
18 September 1980 7 days 20 h 43 min
Yuri Romanenko, Arnaldo Mendez

Yet another Intercosmos mission, this time with a cosmonaut researcher from Cuba – the first coloured spaceman who at 13 years had been a shoeshine boy. One experiment carried out involved the study of the crystallization of sucrose in zero g for the Cuban sugar industry.

Flight 77. (USSR 46) Soyuz T3
27 November 1980 12 days 19 h 8 min
Leonid Kizim, Oleg Makarov,
Gennadi Strekalov

For the first time since June 1971, three Russians lifted off together in a Soyuz. This crew's job was to carry out major maintenance work on Salyut 6 before the launch of a new long-duration crew, including an overhaul of the hydraulic, control refuelling and communications systems.

Flight 78. (USSR 47) Soyuz T4
12 March 1981 74 days 17 h 38 min
Vladimir Kovalyonok, Viktor Savinykh

Cosmonaut Savinykh had the distinction of being not only the 50th cosmonaut but also the 100th person in space during this relatively short mission in Salyut 6, probably on its last legs, and only to facilitate the hosting of two more Intercosmos crews. Savinykh was an expert in aerial photography and cartography.

Flight 79. (USSR 48) Soyuz 39
22 March 1981 7 days 20 h 43 min
Vladimir Dzhanibekov,
Jugderdemidyin Gurragcha

Mongolia's cosmonaut with the near-unpronounceable christian name, was a member of the penultimate Soviet bloc Intercosmos crew. Soon after the return of this crew, on 25 April, Russia launched Cosmos 1267 which docked with Salyut 6 on 19 June 1981 and remained attached to it until July 1982 when the station re-entered. This was the first Star Module, a craft almost as large as Salyut itself and seen as an important part of modular space stations of the future.

Flight 81. (USSR 49) Soyuz 40
15 May 1981 7 days 20 h 38 min
Leonid Popov, Dumitru Prunariu

This was the final mission to Salyut 6 and the final flight by the old-style Soyuz. By the end of this crew's stay and that of the resident Soyuz T4 residents, Salyut had received 16 crews and 15 unmanned craft during 676 days of manual occupation.

Flight 82. (USA 33) STS 2
12 November 1981 2 days 6 h 13 min 11 s
Joe Engle, Richard Truly

Another much delayed flight of the *Columbia* shuttle, originally planned for launch on 30 September, was nonetheless the first flight in history of a used spacecraft. A fuel cell failure caused a reduction in flight time from 5 to 2 days but the crew claimed to have attained 80 per cent of their flight plan, including the testing of a Remote Manipulator System mechanical arm.

Flight 83. (USA 34) STS 3
22 March 1982 8 days 0 h 4 min 46 s
Jack Lousma, Gordon Fullerton

Although minor failures plagued this mission and were highlighted by the press, *Columbia*'s third test flight was a remarkable success, carrying the largest load of experiments and test packages by far. Because Edwards Air Force Base was waterlogged, the landing was scheduled at White Sands, New Mexico on the seventh day but high winds gave Lousma and Fullerton an extra day in space.

Dumitru Prunariu and Leonid Popov, right next to their capsule shortly after landing. Note they have autographed Soyuz with chalk, a traditional ceremony performed by returning cosmonauts. (*Novosti*)

Flight 84. (USSR 50) Soyuz T5
13 May 1982 211 days 8 h 5 min
(landed in Soyuz T7) Anatoli Berezovoi, Valentin Lebedev

This record-breaking long-duration crew hosted a French Intercosmos team and the first lady since Tereshkova to enter space, and also hand-launched two communications satellites called Iskra, from the airlock of Salyut 7, the newly arrived base in space, launched on 19 April 1982. The crew carried out an EVA on 30 July to disassemble and partially replace scientific equipment and to retrieve samples. During their stay Progress 13, 14, 15 and 16 were launched on resupply missions. Lebedev felt lonely and depressed during his stay and paradoxically resented the visitors of T6 and T7 crews which 'spoiled the relationship' he had built up with Berezovoi.

The first men to stay over 200 days in space were the Soyuz T5 visitors to Salyut 7. Anatoli Berezovoi, left, and Valentin Lebedev. (*Novosti*)

Launch of STS 4. (*NASA*)

Flight 85. (USSR 51) Soyuz T6
24 June 1982 7 days 22 h 42 min
Vladimir Dzhanibekov, Alexander Ivanchenkov, Jean-Loup Chrétien

The 51st Soviet-manned space flight marked the launch of the first Western European in space. Frenchman Chrétien criticized the flight plan after his mission because he felt that he and his crew were kept far too busy. He also remarked that he found the re-entry and landing far more dramatic than the launch.

Flight 86. (USA 35) STS 4
27 June 1982 7 days 1 h 9 min 31 s
Ken Mattingly, Hank Hartsfield

The fourth and final test flight of shuttle *Columbia* marked the first time in a US-manned space flight that military payloads were flying. Much of the television coverage and conversations from the crew were restricted for security reasons. One payload, called Cirris, a cryogenic infra-red radiance instrument to obtain spectral data on the exhausts of vehicles powered by rocket and air breathing engines, did not work because its lens cap failed to come off.

Flight 87. (USSR 52) Soyuz T7
19 August 1982 7 days 21 h 52 min
(landed in Soyuz T5) Leonid Popov, Alexander Serebrov, Svetlana Savitskaya

The first flight of a woman into space for almost 20 years was linked to the fact that Sally Ride had recently been named to be America's first space lady on shuttle mission 7. Whereas Tereshkova, the first woman in space in 1963, was a cotton-mill worker and amateur parachutist, Savitskaya was a highly qualified test pilot and holder of 18 aviation world records.

Flight 88. (USA 36) STS 5
11 November 1982 5 days 2 h 14 min 26 s
Vance Brand, Robert Overmyer, Joseph Allen, William Lenoir

This flight of *Columbia* was the first commercial venture for the space shuttle. Two communications satellites, SBS-C and Anik C-3, were successfully deployed into orbit. However, a planned spacewalk by Allen and Lenoir had to be cancelled due to faulty equipment in the Portable Life Support System backpacks. The flight also marked the first by four men in space at once.

Flight 89. (USA 37) STS 6
4 April 1983 5 days 0 h 23 min 42 s
Paul Weitz, Karol Bobko, Story Musgrave, Donald Peterson

The maiden flight of shuttle orbiter *Challenger* had to be delayed for three months due to serious engine faults but the launch on 4 April was perfect, as was the deployment of the TDRS satellite NASA's tracking station in the sky – and a spacewalk by Musgrave and Peterson. But an IUS upper stage failure, which stranded TDRS in the wrong orbit, spoiled the flight.

The maiden flight of Space Shuttle *Challenger* ends with a landing at Edwards Air Force Base. It was the sixth shuttle mission. (*NASA*)

Flight 90. (USSR 53) Soyuz T8
20 April 1983 2 days 0 h 20 min
Vladimir Titov, Gennadi Strekalov, Alexander Serebrov

The rendezvous radar antenna on Soyuz T8 failed to deploy on this mission to place a possible maintenance crew on Salyut 7. Despite this, Titov did try a docking but 525 feet away he realised that he was travelling too fast for safety and backed off.

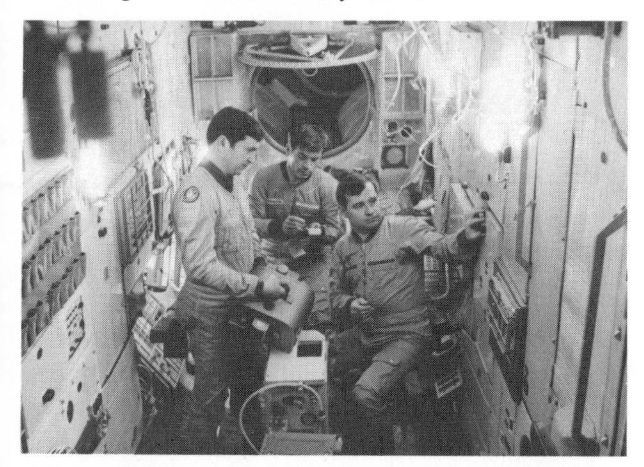

This should have been the scene inside Salyut 7 had Soyuz T8 made a successful docking. In a Salyut mock-up are the Soyuz T8 crew, left to right, Vladimir Titov, Alexander Serebrov and Gennadi Strekalov. (*Novosti*)

Flight 91. (USA 38) STS 7
18 June 1983 6 days 2 h 24 min 10 s
Robert Crippen, Rick Hauck, Sally Ride, John Fabian, Norman Thagard

The first five person space crew, on *Challenger*, included Sally Ride, the first US lady in space. STS 7 deployed two comsats, Anik C2 and Palapa B, and one satellite called SPAS, a German-built science package, which was retrieved – another space first – by the Remote Manipulator System. *Challenger*'s planned first-time landing at Cape Canaveral had to be cancelled due to bad weather, so it came home to Edwards Air Force Base.

Flight 92. (USSR 54) Soyuz T9
27 June 1983 149 days 9 h 46 min
Vladimir Lyakhov, Aleksander Aleksandrov

Clearly, what was to have been a flight to exceed the T5 crew's 211 days, this mission was bravely endured by a crew who had to contend with a mountain of problems with Salyut 7, not the least of which was a leaking propulsion system and a descent craft rapidly exceeding its safe lifetime in space. A new Star Module, Cosmos 1443, docked with Salyut, plus Progress 17 and 18, but not a fresh craft – Soyuz T10 aborted on the pad. Lyakhov and Aleksandrov went for two spacewalks to add new solar panels to the space station but came home on 23 November.

Launch of shuttle mission 8 is the second night launch in the US space programme. (*NASA*)

Flight 93. (USA 39) STS 8

30 August 1983 6 days 1 h 8 min 40 s
Richard Truly, Daniel Brandenstein, Dale Gardner, William Thornton, Guion Bluford

This crew, which included the oldest man in space Bill Thornton, fifty-four, and the first black astronaut, Guion Bluford, was the first to be launched from the US at night since Apollo 17. The mission deployed the comsat Insat and tested the capability of the RMS to handle heavy payloads using a Payload Flight Test Article, PTFA. A night landing was made for the first time in the US space programme at Edwards.

SOYUZ T10-1

27 September 1983 Vladimir Titov and Gennadi Strekalov

An attempt to launch a pair of maintenance men to the half-crippled Salyut 7 and to provide its resident crew with a fresh descent craft failed on 27 September 1983, during the first launch-pad explosion in manned space flight history. Fortunately, a launch escape system hauled Titov and Strekalov inside Soyuz free of the conflagration and they landed 2 miles away, shaken and in need of a stiff vodka!

Flight 94. (USA 40) STS 9

28 November 1983 10 days 7 h 47 min 23 s
John Young, Brewster Shaw, Owen Garriott, Robert Parker, Ulf Merbold, Byron Lichtenberg

The first flight of the European Space Agency's Spacelab was also the first with six men in one craft and the first in which a new breed of astronaut-passenger, the payload specialist was to fly. Young was making his sixth flight. The mission was hailed as a success and was extended one day. A scientific bonanza resulted from the over 70 detailed experiment packages carried in the Spacelab module inside *Columbia*'s payload bay.

Flight 95. (USA 41) STS 41B

3 February 1984 7 days 23 h 15 min 54 s
Vance Brand, Robert Gibson, Bruce McCandless, Robert Stewart, Ronald McNair

Although *Challenger*, on the tenth shuttle mission and saddled with a new, confusing numbering system, successfully deployed two comsats, Westar and Palapa, the failure of their upper stages was blamed on the shuttle by ill-informed sections of the popular press. This had a negative effect on the programme's image just when it was getting over its early problems. The highlight of the mission however, was the untethered spacewalk of Bruce McCandless wearing the first manned manoeuvring unit. *Challenger* came home to Cape Canaveral, the first time a craft had landed at its launch site. McCandless had the longest wait so far to get into space since selection as an astronaut – 18 years.

Flight 96. (USSR 55) Soyuz T10

8 February 1984 236 days 22 h 50 min
Leonid Kizim, Vladimir Solovyov, Oleg Atkov

This record-breaking flight featured three crewmen, not two; and one of them was a cardiologist, Dr Atkov, the first of a team of cosmonaut doctors who were to fly all future long duration missions. A total of 87 days was spent on medical examinations during the flight. Kizim and Solovyov carried out a record of six spacewalks, totalling almost a day, to assemble new solar power panels outside the station. The crew also hosted the final Intercosmos cosmonaut from India and Savitskaya on her second flight.

Flight 97. (USSR 56) Soyuz T11

3 April 1984 7 days 21 h 41 min
Yuri Malyshev, Gennadi Strekalov, Rakesh Sharma

The final Intercosmos flight was made by Indian cosmonaut Sharma, who became the third most famous person in his country behind Mrs Gandhi and Amitabh Bachchan, a film star. For the first time, six people were on board Salyut at one time and, with STS 41C in orbit, a record of 11 people were in space at one time.

Flight 98. (USA 42) STS 41C

6 April 1984 6 days 23 h 40 min 5 s
Robert Crippen, Francis Scobee, Terry Hart, James van Hoften, George Nelson

Challenger deployed a passive satellite called Long Duration Exposure Facility, a 12-sided cylinder containing 57 experiments, which was to remain in orbit to be retrieved on a later flight. Then a rendezvous was made with the ailing satellite Solar Maximum Mission or Solar Max. Nelson, wearing an MMU, flew out to Solar Max with a view to stabilizing it for retrieval. He failed but the RMS managed to grab it. Nelson and van Hoften repaired Solar Max during an EVA. The satellite was then deployed.

Flight 99. (USSR 57) Soyuz T12

17 July 1984 11 days 19 h 14 min
Vladimir Dzhanibekov,
Svetlana Savitskaya, Igor Volk

Svetlana Savitskaya beat Sally Ride into space in 1982 and – just before Ride was to become the first lady to fly in space twice and Kathy Sullivan, her companion on shuttle 41G, the first woman to walk in space – Savitskaya did it again. During their longer than usual stay for a visiting crew, Savitskaya and Dzhanibekov performed an EVA, the space lady wielding a portable electron welding torch to carry out cutting, welding and soldering tests in space.

Flight 100. (USA 43) STS 41D

30 August 1984 6 days 0 h 56 min 4 s
Hank Hartsfield, Michael Coats,
Steven Hawley, Richard Mullane,
Judith Resnik, Charles Walker

The launch of *Discovery*, a new shuttle orbiter, was delayed two months by an engine failure during an attempted blast-off on 27 June – the second US launch-pad abort. At the fourth attempt, it reached orbit and deployed three comsats, erected a new solar sail device to provide additional electrical power, while industry-astronaut Charlie Walker operated an Electrophoresis unit to manufacture a hormone in zero g which could form the basis of a new pharmaceutical. 41D was the heaviest total manned spacecraft in orbit weighing 263 000 lb.

Flight 101. (USA 44) STS 41G

5 October 1984 8 days 5 h 23 min 33 s
Robert Crippen, Jon McBride, Sally Ride,
Kathy Sullivan, David Leestma,
Paul Scully-Power, Marc Garneau

The first seven-person flight in history, *Challenger*'s sixth mission featured the deployment of Earth Radiation Budget Satellite, a spacewalk by Sullivan – the first US woman – and Leestma to demonstrate a technique for orbital refuelling, the work of oceanographer Scully-Power and Canada's first astronaut Garneau.

Flight 102. (USA 45) STS 51A

8 November 1984 7 days 23 h 45 min 54 s
Frederick Hauck, David Walker, Joseph
Allen, Dale Gardner, Anna Fisher

The Westar and Palapa satellites, launched by STS 41B, were stranded in useless orbits by failures of their own rocket motors and during this mission they were retrieved and brought back to Earth in one of the most spectacular manned space flights in history. EVA crewmen Gardner and Allen manhandled spacecraft weighing on Earth over 1000 lb. STS 51A also deployed two of its own comsats.

Flight 103. (USA 46) STS 51C

24 January 1985 3 days 1 h 33 min 13 s
Ken Mattingly, Loren Shriver, Elison
Onizuka, James Buchli, Gary Payton

The USA's first classified military manned space flight was preceded with enough publicity to declassify it to such an extent that the nature of orbiter *Discovery*'s payload was well known. An electronic monitoring satellite was deployed into orbit attached to its IUS upper stage. The USAF's first Manned Space Flight Engineer, Gary Payton, flew as payload specialist.

Flight 104. (USA 47) STS 51D

12 April 1985 6 days 23 h 56 min
Karol Bobko, Don Williams, Rhea
Seddon, Jeff Hoffman, David Griggs,
Charles Walker, Jake Garn

The NASA crew on this flight were originally going to fly in August 1984, but their missions were scrapped twice and they were reassigned to 51D, a controversial flight because chosen as a payload specialist was US Senator Jake Garn who just happened to be the Chairman of a Congressional committee that oversees NASA's budget. Some critics called Garn's space trip 'the ultimate junket' but his flight was not criticized universally, although it became the butt of many jokes. Again, inaccurate but perhaps inevitable reporting in the press blamed the shuttle for losing one of the satellites it in fact successfully deployed. A brave attempt to mend the Leasat failed. The flight featured an unscheduled EVA and a second mission by payload specialist Charlie Walker.

Flight 105. (USA 48) STS 51B/Spacelab 3

29 April 1985 7 days 0 h 8 min 50 s
Robert Overmayer, Fred Gregory,
Don Lind, William Thornton,
Norman Thagard, Lodewijk van den Berg,
Taylor Wang

This flight of the Spacelab 3 science laboratory featured 15 experiments in materials processing, fluids mechanics, life sciences, atmospheric physics and astronomy. With the 7 man crew were 24 rats and 2 monkeys. The crew included three men over 50 years old; the oldest man in space again, Bill Thornton at 56 and Don Lind, flying his first mission after a record wait of 19 years since selection as an astronaut, and a payload specialist Lodewijk van den Berg. Despite niggling technical problems which were as usual highlighted in the press, all but one of the experiments yielded excellent results.

2

Firsts in Manned Space Flight

The first aborted ascent into space took place on the launch of Soyuz 18-1 on 5 April 1975. Although its launcher's second stage engine ignited, the stage had not separated from the centre core of the first stage. Soyuz was ejected and made a sub-orbital spaceflight landing 200 miles from China. Cosmonauts were Lazarev and Makarov.

Oleg Makarov, left, and Vasili Lazarev, victims of the Soyuz 18-1 abort. (*Novosti*)

The first aborted launch of a manned spacecraft on the launch pad after ignition of its booster was Gemini 6 (Schirra and Stafford) on 12 December 1965 which was within a split second of lift-off when the Titan 2 engines shut down after detecting a fault. A similar known launch-pad abort did not take place again until STS 41D on 27 June 1984.

The Gemini 6 launch pad abort. (*NASA*)

The first spacecraft to make alterations to its orbit was Gemini 3, launched on 23 March 1965. The spacecraft, manned by Gus Grissom and John Young, was also equipped with the first computer.

A view from Gemini 3. (*NASA*)

The Baikonur Cosmodrome, 12 April 1961. The beginning. Vostok 1 is launched and inside a cannonball-shaped craft is Yuri Gagarin, the first man in space. (Note: the Cosmodrome is actually 250 miles from Baikonur and in the West is referred to as Tyuratam, its actual location.) (*Novosti*)

Cape Canaveral, 5 May 1961. America's reply. Freedom 7, a Mercury capsule, is launched by a Redstone rocket. Inside is Alan Shepard, the first astronaut. Shortly after this flight, President Kennedy launched America on course for the Moon. Project Apollo and the 'Moon Race' dominated the sixties and overshadowed many applications of space that are only being appreciated today. (*NASA*)

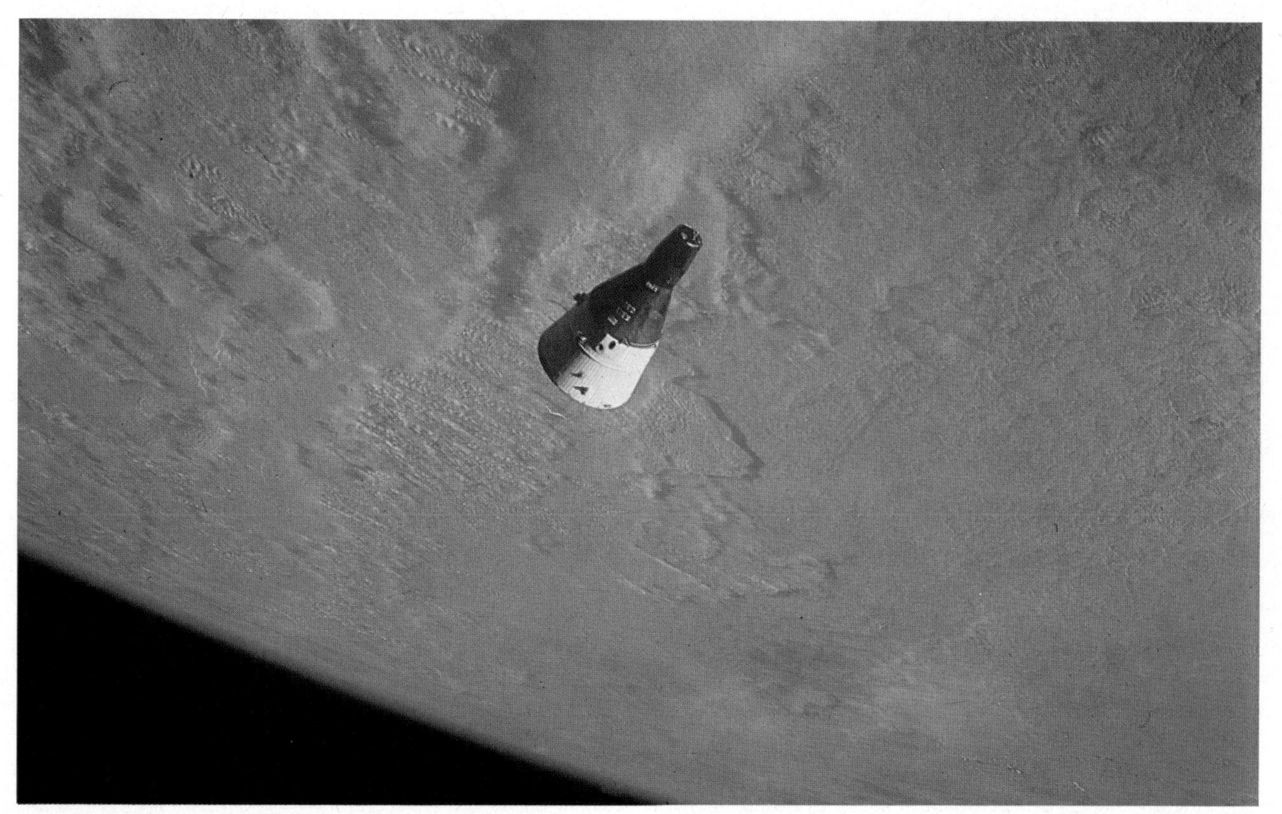

Earth orbit, 15 December 1965. Rendezvous. A major milestone on the road to the Moon was the first rendezvous in space made by Gemini 6, which flew to within 6 inches of Gemini 7 already in orbit, simulating the flight of the Apollo lunar module ascent stage from the Moon to the command module orbiting it. (*NASA*)

Edwards Air Force Base, California, 1968. The unofficial spacemen. While Gemini was catching the headlines, the X-15 rocket plane was demonstrating space flight with a reusable vehicle. Thirteen flights of the X-15, between 1962 and 1968, exceeded a height of 50 miles and although not officially recognized as space flights, they are worthy of record in any history of manned space flight. There were twelve X-15 pilots and eight of them flew 'astroflights'. The pilots here are, left to right, Joe Engle (now a Shuttle commander), Milton Thompson, Robert Rushworth, Walt Williams (project executive), Pete Knight, the late Jack McKay and Bill Dana. (*NASA*)

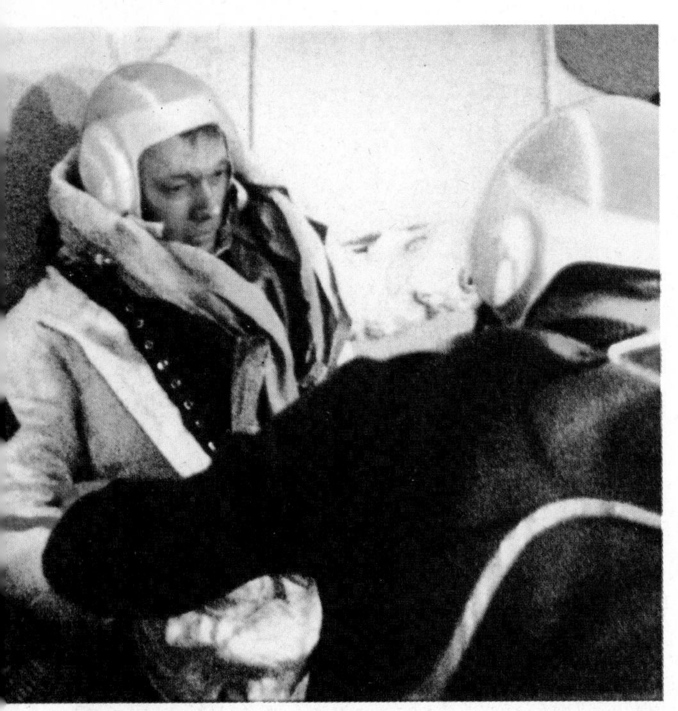

Inside Soyuz 5, 16 January 1969. Preparing for a spacewalk. Alexei Yeliseyev dons his spacesuit in preparation for the world's first crew transfer between spacecraft. Soyuz 5 had docked with Soyuz 4 and Yeliseyev and his companion, Yevgeni Khrunov, made a spacewalk to Soyuz 4 in which they returned to Earth. As the USA pressed on to the Moon, the Soviets were preparing for the first flights to space stations. (*Novosti*)

The Tyuratam Cosmodrome, 1 June 1970. Soyuz 9 prepares for flight. Inside this spacecraft two cosmonauts lived in space for 17 days. They were in such bad shape when they landed that they had to be carried from the spacecraft. Long term weightlessness had taken its toll, particularly on their ability to readapt to gravity. Thanks to strenuous exercise routines for future cosmonauts, the record now stands at 236 days. (*Novosti*)

The first American to fly inside a Russian spacecraft was Vance Brand who was first to visit the Soyuz cabin during the ASTP flight launched on 15 July 1975.

Inside Soyuz during training for ASTP are, left to right, Alexei Leonov, Vance Brand and Valeri Kubsaov. (*Novosti*)

The first manned spaceflight to create artificial gravity in orbit was Gemini 11, which was launched on 11 September 1966. Gemini flew in spinning formation with Agena 11, which was attached to Gemini by a 100 ft tether.

The first male–female spacewalk was made by Vladimir Dzhanibekov and Svetlana Savitskaya, launched on 17 July 1984. The first by Americans was made by David Leestma and Kathryn Sullivan on STS 41G.

The first and so far only person to clock up over one day's EVA experience is American Eugene Cernan, who with walks from Gemini 9 in June 1966 and Apollo 17 in December 1972 has 24 h 12 min EVA time to his credit.

The first and so far only Soviet-manned space launch shown live across the world was that of Soyuz 19 on 15 July 1975 on its mission to dock with a US Apollo.

The first person over fifty to fly into space was Deke Slayton who, at fifty-one, was docking module pilot of Apollo 18/ASTP launched on 15 July 1975.

The first men to spend a day on the Moon were Charles Conrad and Alan Bean of Apollo 12 launched on 14 November 1969. The first men to spend two days on the Moon were David Scott and James Irwin of Apollo 15 launched on 26 July 1971. The first and only men to spend three days on the Moon were Eugene Cernan and Jack Schmitt of Apollo 17 launched on 7 December 1972.

The first man to dock manually with a satellite was George Nelson, mission specialist of STS 41C, launched on 6 April 1984, who used a 'T-PAD' docking device to 'soft-dock' with the Solar Max satellite. The first to 'hard-dock', this time using a 'stinger', was Joe Allen who latched onto Palapa in November 1984 on mission 51A.

The first person to walk in space twice was Edwin Aldrin who performed an EVA on Gemini 9 and walked on the Moon during the Apollo 11 mission launched on 16 July 1969. The first Russian was Vladimir Lyakhov who walked in space on Soyuz 32 and Soyuz T9.

The first person to walk in space three times was David Scott who made three lunar EVAs on Apollo 15, launched 26 July 1971. The first Russian was Vladimir Lyakhov during his second EVA on mission Soyuz T9.

The first person to walk in space four times was Eugene Cernan who performed an EVA in 1966 on Gemini 9 and three Moon walks during the Apollo 17 mission launched on 7 December 1972.

The first persons to make five and six spacewalks – and four in Earth orbit – were Leonid Kizim and Vladimir Solovyov on Soyuz T10 on 8 February 1984.

The first person to fly a space mission when he had already been named to fly a subsequent mission was Robert Crippen who commanded STS 7 when already assigned to 41C (a mission he also flew when already assigned to 41G).

Deke Slayton, in space at 51. (*NASA*)

The first flight to carry a space-experienced multi-crew was Apollo 10, launched on 18 May 1969 and crewed by Stafford, Young and Cernan. The first Soviet space-experienced multi-crew flight was Soyuz 8's Shatalov and Yelisyev.

The first manned spaceflight to be boosted by solid fuel rockets was STS 1, launched on 12 April 1981, powered by two Solid Rocket Boosters, augmenting the thrust of the orbiter's SSMEs.

The first manned spaceflight to be flown by a replacement astronaut was MA-7 on 24 May 1962 and piloted by Scott Carpenter who replaced Deke Slayton, who was diagnosed to have a heart flutter. Slayton's back-up Schirra was not selected. The first manned mission flown by a back-up crew was Gemini 9, launched in June 1966. Tom Stafford and Eugene Cernan were the original back-up crewmen who replaced the prime crew killed in an air crash in February 1966.

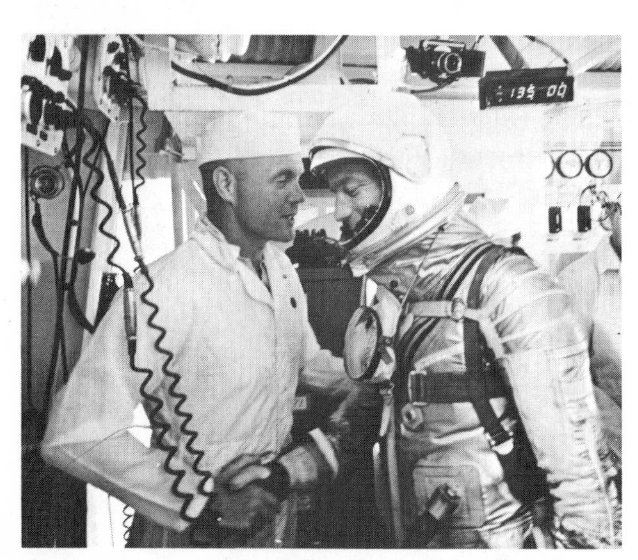

John Glenn, left, wishes good luck to Scott Carpenter before MA7. (*NASA*)

The first astronaut to be launched on his birthday was STS 2 pilot Richard Truly on 12 November 1981. The second was STS 51A mission specialist, Dale Gardner on 8 November 1984.

The first manned spacecraft to be launched and to land at night was Soyuz 10 which went aloft on 23 April 1971.

The first bachelor in space was Vostok 3 pilot Andrian Nikolyev, launched on 11 August 1962 who in 1964 married the first spinster Valentina Tereshkova, Vostok 6 pilot, launched on 16 June 1963. They are now divorced. The first US bachelor spaceman was Apollo 13 CMP, the late Jack Swigert. The first and so far only bachelor on the Moon was Apollo 17 LMP Jack Schmitt. The first unmarried US woman in space was divorcee Judy Resnik of STS 41D while Kathy Sullivan of STS 41G is unmarried.

Andrian Nikolyev greets Valentina Tereshkova at the launch pad prior to the first space flight by a woman. (*Novosti*)

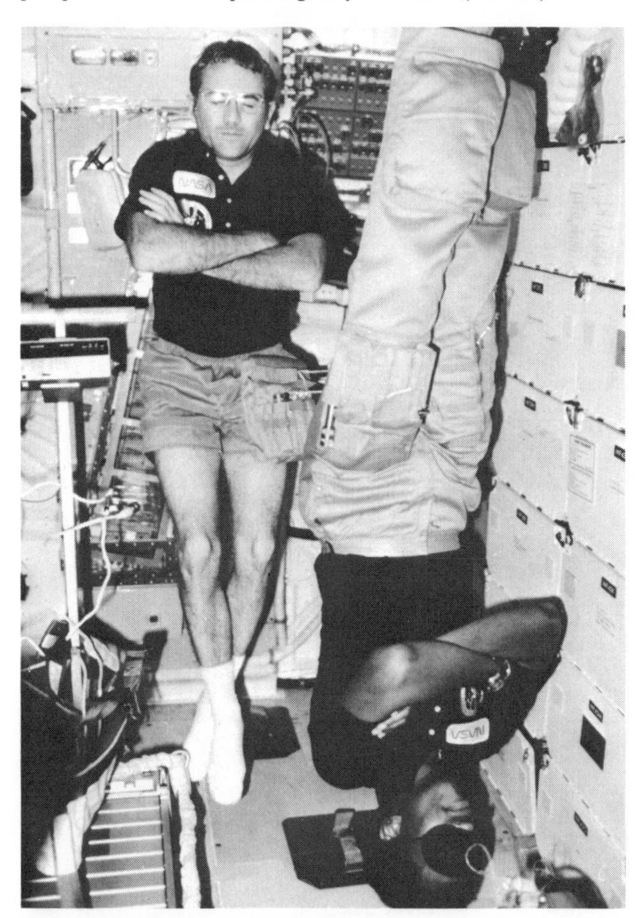

Richard Truly and Guion Bluford asleep during the STS 8 mission. (*NASA*)

The first person to make bodily contact with another craft in orbit was Michael Collins of Gemini 10, launched on 18 July 1966, who 'flew over' to Agena 8 during a curtailed 39 minute EVA at the end of a 50 ft-long tether. There he retrieved samples from the side. The next person to perform such a feat in Earth orbit was George Nelson, mission specialist of STS 41C, launched 6 April 1984 during his EVA to retrieve Solar Max satellite. (Apollo 12's lunar landers Conrad and Bean made contact with Surveyor 3 on the Moon in November 1969.)

Mike Collins leaves Gemini 10 after splashdown. (*NASA*)

The first spacewalk by a spaceman in cislunar space – that is, between the Moon and the Earth – took place during the return journey from the Moon of Apollo 15 launched on 26 July 1971. It was made by Major Alfred Worden, USAF, 199 000 miles from Earth, travelling at a speed of 2000 mph. His EVA was televised and lasted 34 minutes. Similar EVAs were performed by Ken Mattingly and Ronald Evans on Apollo 16 and 17 respectively in 1972.

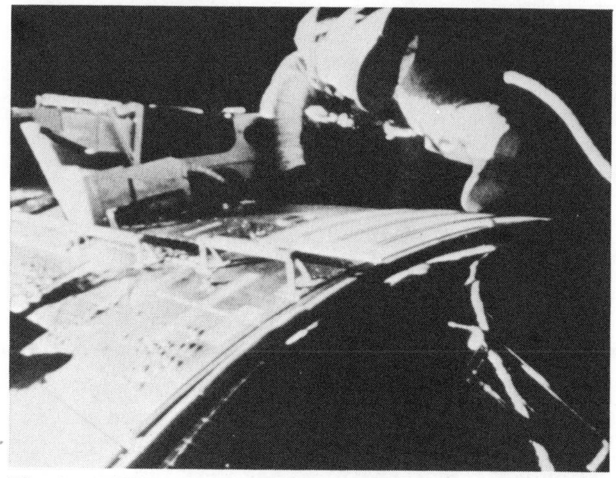

The first TEEVA by Al Worden. (*NASA*)

The first civilian commander of a space mission was Neil Armstrong who led the flight of Gemini 8 on 16 March 1966. He was also the first US civilian in space and the first of two to walk on the Moon. The first Soviet civilian commander was Nikolai Ruchavishnikov of Soyuz 33.

The first civilians in space were Konstantin Feoktistov and Boris Yegerov, who flew in Voskhod 1 on 12 October 1964. They were also the first scientist and first doctor in space.

The first hands-off computer-controlled re-entry and landing was that of Gemini 11, launched on 12 September 1966. Of course, Soviet Vostoks were brought home under remote control from the ground.

The first spaceman to fly consecutive spaceflights for his country was Alexander Serebrov, flight engineer of Soyuz T8, launched on 20 April 1983. He was also the flight engineer of Soyuz T7 in 1982.

Alexander Serebrov. (*Novosti*)

The first crew transfer to take place in space occurred when Soyuz 5 crewmen Alexei Yeliseyev and Yevgeny Khrunov made the first dual spacewalk to the docked Soyuz 4. They became the first people to land in a craft that they were not launched in. Soyuz 5 was launched on 15 January 1969.

Alexei Yeleseyev, left and Yevgeni Khrunov. (*Novosti*)

The first manned flight to be curtailed was the Gemini 5 launched on 21 August 1965 which landed one orbit early to miss hurricane Betsy. The first to be curtailed due to an emergency was Gemini 8 on 16 March 1966. It went out of control due to a short-circuited thruster firing continuously.

The first deployment of an operational satellite from a manned craft took place during the mission of Apollo 15, launched on 26 July 1971 when the Apollo subsatellite was ejected in lunar orbit. Iskra comsats have been hand-launched from Salyut 6 since then and many satellites from the shuttle.

The first deployment of a satellite from a manned spacecraft was made by Major Gordon Cooper, the pilot of MA 9 *Faith 7*, on 15 May 1963, who ejected a 6 inch diameter flashing beacon from the capsule to enable him to test his ability to perceive distances and visibility of objects in space.

Gordon Cooper. (*NASA*)

The first spaceman to die of natural causes was Pavel Belyayev, commander of Voskhod 2, who died on 10 January 1970 following an operation for stomach ulcers. US astronaut Apollo 13 CMP Jack Swigert died of cancer in December 1982.

Pavel Belyayev before his launch on Voskhod 2. (*Novosti*)

The first man to die on a space mission was Vladimir Komarov of Soyuz 1 on 24 April 1967. The Soviets say that the single parachute on Soyuz snarled at 25 000 ft and the craft plummeted to the ground. However, there is evidence to suggest that the cosmonaut was in trouble before this. The three crewmen of Soyuz 11 died during their re-entry on 29 June 1971 and three astronauts died on the ground in the Apollo 1 fire on 27 January 1967.

The late Vladimir Komarov. (*Novosti*)

The first manned space docking took place on 16 March 1966 when Gemini 8, crewed by Armstrong and Scott, joined up with an Agena 8 unmanned target rocket. The craft remained together for 27 minutes before a malfunction forced Gemini 8 to return to Earth early.

The first and so far only docking and joint flight involving the Soviet Union and USA was the Apollo Soyuz Test Project. On 17 July 1975 over Amsterdam, Apollo 18 – manned by Stafford, Slayton and Brand – docked with Soyuz 19, manned by Leonov and Kubasov.

The first dual flights by manned spacecraft were made by Vostok 3 and 4 on 12 August 1962, the day the latter spacecraft was launched. The craft passed within four miles of each other in what was wrongly claimed by the Russians to be the first rendezvous in space.

The first escape from an exploding rocket during launch was made by cosmonauts Titov and Strekalov of Soyuz T10-1 on 27 September 1983. When the A2 launcher caught fire shortly before ignition, the launch escape rocket system hauled Soyuz away from the conflagration. The crew was shaken but not hurt.

The launch escape system that saved the lives of the Soyuz T10-1 crew. (*Novosti*)

The first manned spaceflight to be extended was Voskhod 2, launched on 18 March 1965. Its automatic retrofire sequence failed on orbit 17 and commander Belyayev had to perform a manual retrofire on orbit 18. The spacecraft landed 500 miles from its intended target, in a forest near Perm. The first extended US flight was Apollo 9 in March 1969 which flew one extra orbit to miss heavy seas on splashdown.

The first spaceman to become a father while in space was Leonid Kizim during his 237 day Soyuz T10/Salyut 7 flight launched on 8 February 1984.

The first fare-paying passenger in space was Charles Walker for whom his company, McDonnell Douglas, paid NASA £40 000 to fly him on STS 41D on 30 August 1984, to operate their electrophoresis experiment.

Charlie Walker. (*McDonnell Douglas*)

The first flight of over 50 days was made by astronauts Alan Bean, Owen Garriott and Jack Lousma, launched to Skylab on 28 July 1973.

The first man in space was Major Yuri Gagarin of the Soviet Air Force, who was pilot of Vostok 1 on 12 April, 1961. He was also the first man to land separately from his spacecraft – a fact not admitted by the Russians until 1978.

Gagarin preparing to leave the Earth. (*Novosti*)

The first four-man space launch was that of STS 5-*Columbia* on 11 November 1982 carrying Vance Brand, Robert Overmyer, Joe Allen and William Lenoir.

The first man in space four times was James Lovell when he was launched on Apollo 13 on 11 April 1970, having previously flown Gemini 7, Gemini 12 and Apollo 8. The first Russian to make four flights is Oleg Makarov (Soyuz 12, Soyuz 18-1, Soyuz 27 and Soyuz T3).

James Lovell, left, the first man to make four space missions and Buzz Aldrin, the first to make two EVAs. (*NASA*)

The first five-person space launch was STS 7, which took off on 18 June 1983 with Robert Crippen, Rick Hauck, John Fabian, Sally Ride and Norman Thagard aboard.

The first man to make both five and six spaceflights was John Young. When he commanded the first space shuttle mission, STS 1 on 12 April 1981 he had flown five times. His previous flights were Gemini 3, Gemini 10, Apollo 10 and Apollo 16. Then on 28 November 1983 he commanded STS 9-Spacelab 1, his sixth spaceflight.

John Young, foreground, suiting up for STS 1, with Bob Crippen. (*NASA*)

The first grandfather in space was Lev Demin, forty-seven, flight engineer of Soyuz 15, launched on 26 August 1975.

Grandfather Lev Demin. (*Novosti*)

The first man to fly in space independently of his spacecraft was Bruce McCandless on the STS 41B mission, launched on 3 February 1984. He was the first man to operate a manned manoeuvring unit. Other astronauts to have used MMUs are Robert Stewart (41B), George Nelson and James van Hoften (41C) and Joe Allen and Dale Gardner (51A). (Yuri Romanenko floated untethered a short way outside Salyut 6 in 1978 before being grabbed by his companion.)

The first manned spacecraft to land at its launch site was STS 41B, which went into orbit on 3 February 1984 and came home to Cape Canaveral eight days later.

The first spacecraft to be lost was *Liberty Bell 7*, which sank under the Atlantic Ocean at the end of its flight, on 21 July 1961. Its pilot Gus Grissom was saved from drowning at the end of the third manned space flight.

The first manned lunar roving vehicle was LRV 15, which was deployed on the Moon by the crew of Apollo 15, launched on 26 July 1971. Dave Scott and James Irwin drove 17.3 miles in the LRV at Hadley Rille during three EVAs. LRVs were also driven by the crews of Apollo 16 and 17.

Bruce McCandless, a human satellite. (*NASA*)

The first dedicated manned military spying mission was Soyuz 14 launched on 3 July 1974 to Salyut 3. The pilots were Pavel Popovich and Yuri Artyukin. The first full US military mission was STS 51C in January 1985 but STS 4 did carry a military payload in June 1982.

Pavel Popovich, left, and Yuri Artyukhin. (*Novosti*)

The first man to manoeuvre his spacecraft was Commander Alan Shepard pilot of the second manned spaceflight, *Freedom 7* on 5 May 1961. Using gas thrusters, he performed pitch, yaw and roll movements. He was also the first spaceman to come down at sea.

The first flight to the Moon and to achieve escape velocity was made by Apollo 8 on 21 December 1968. The astronauts Frank Borman, James Lovell and Williams Anders flew ten lunar orbits over Christmas.

The historic ascent of Apollo 8. (*NASA*)

The first men on the Moon were Neil Armstrong and Buzz Aldrin who landed at Tranquillity Base at 9.18 pm BST on 20 July 1969. Armstrong walked on the Moon at 3.56 am BST on 21 July 1969. His first words were 'That's one small step for Man, one giant leap for mankind'. He did not in fact say 'a man'.

Neil Armstrong, left, and Buzz Aldrin, ten years after. (*NASA*)

The first mother in space – and the first female medical doctor – was Anna Fisher, mission specialist of the satellite retrieval mission STS 51A, launched on 8 November 1984. (She was the sixth woman in space and fourth American.)

The first Negro spaceman was Guion Bluford, a mission specialist on STS 8, launched on 30 August 1983. The first coloured spacemen was Arnaldo Mendez of Soyuz 38 in 1980. The second US negro astronaut was Ronald McNair of STS 41B in 1984.

The first spaceman to come from neither the USA nor the Soviet Union was Czechoslovakian research pilot of Soyuz 28 on a mission to Salyut 6, Captain Vladimir Remek. Soyuz 28 which was launched on 2 March 1978 was the first of 11 Intercosmos flights involving pilots from other Soviet Bloc countries and also France and India.

The first non-Soviet-American spaceman, Vladimir Remek (Czechoslovakia), right, at the launch pad with Alexei Gubarev. (*Novosti*)

The first night launch of a manned space flight was that of Soyuz 1 at 3.35 am on 23 April 1967 carrying Vladimir Komarov. Since then the Soviets have launched many of their Soyuz at night. The first US night launch was Apollo 17 on 7 December 1972. The next was the launch of the shuttle STS 8 on 30 August 1983.

The first night landing by a manned spacecraft was made by Soyuz 10 after its unsuccessful flight to Salyut 1, starting on 23 April 1971. Two days later Soyuz hit the ground at about 2.00 am. The first night landing by a US spacecraft was STS 8 in September 1983.

The first non-pilot commander of a multi-manned spacecraft was Nicholai Ruchavishnikov who led the aborted mission of Soyuz 33 on 10 April 1979. Ruchavishnikov, a spacecraft designer and expert in engineering physics, first flew in 1971. The only other non-pilot commander was Valeri Kubasov of Soyuz 36.

Nikolai Ruchavishnikov. (*Novosti*)

The first non-pilot space person was Valentina Tereshkova, launched in Vostok 6 on 16 June 1973. The first non-pilot US astronaut was Dr Sally Ride, mission specialist on STS 7 in June 1983.

The first manned space flight to exceed 100 days was made by Soyuz 29 cosmonauts Vladimir Kovalyonok and Alexander Ivanchenkov, who were launched in Salyut 6 on 15 June 1978.

The first manned craft to make a rendezvous with a second target vehicle in space was Gemini 10 which rendezvoused with Agena 8 after its docking flight with Agena 10. Gemini 10 was launched on 18 July 1966.

The first rendezvous in space between two manned craft was held on 15 December 1965 when Gemini 6 (Schirra and Stafford) flew up to meet Gemini 7 (Borman and Lovell). The two spacecraft came to within 6 inches of each other and remained together for 5 h 18 min.

The first spacemen to repair a satellite in orbit for redeployment were George Nelson and James van Hoften of STS 41C, launched on 6 April 1984. Other spacemen to have carried out repairs during EVAs in space before this were Charles Conrad and Joseph Kerwin, outside the Skylab space station, and the Apollo 17 astronauts Eugene Cernan and Jack Schmitt, who repaired their lunar rover on the Moon.

George Nelson prepares to dock with Solar Max. (*NASA*)

The first rocket stage engine restart on a manned flight took place on Gemini 10, launched on 18 July 1966, when the Agena 10 engine, to which it was docked, fired to carry astronauts Young and Collins to a record altitude of 474 miles. A similar manoeuvre on Agena II took the crew of Gemini 11 to a height of 850 miles in September 1966.

The first satellite to be retrieved by a manned spacecraft was SPAS 1 which was deployed and retrieved by STS 7, launched on 18 June 1983. The first to be retrieved and deployed was Solar Max by STS 41C, launched on 6 April 1984.

The first time Russians and Americans flew in space at the same time was on 13 December 1973, when Soyuz 13, manned by Klimuk and Lebedev, joined the Slylab 4 mission (Carr, Gibson, Pogue). The craft did not rendezvous or dock.

The first manned flight to bring satellites back to Earth was STS 51A, launched on 8 November 1984 to retrieve and return comsats Palapa and Westar. (STS 7 did bring home SPAS 1 which had been deployed and retrieved from the shuttle.)

Satellite retrieval on shuttle mission 51A. (*NASA*)

The first seven-person space mission was that of the STS 41G, launched on 5 October 1984 and crewed by Robert Crippen, Jon McBride, Sally Ride, Kathryn Sullivan, David Leestma, Paul Scully-Power and Marc Garneau.

The first manned flight of a secondhand spacecraft was STS 2, launched on 12 November 1981. Space shuttle *Columbia* had flown STS 1 and subsequently made three more consecutive missions.

The first six-man space shot STS 9 – Spacelab 1,was launched on 28 November 1983. On board were John Young, Brewster Shaw, Owen Garriott, Robert Parker, Ulf Merbold and Byron Lichtenberg.

The first man to sleep in space – and the first to be sick – was Vostok 2 pilot Gherman Titov who was launched on 6 August 1961 and flew for a day in space. At 25 years old, he was also the youngest person in space and still is today.

The first man to fly solo in lunar orbit and the first to go into lunar orbit twice was John Young, CMP of Apollo 10 (launched on 18 May 1969) and commander of Apollo 16 (launched on 16 April 1972).

Space sleeper Gherman Titov. (*Novosti*)

The first man to fly the space shuttle twice was Robert Crippen on STS 7 on 18 June 1983, having previously flown on STS 1. Crippen was also the first man to fly the shuttle three and four times on missions 41C and 41G, in 1984, and is scheduled for a fifth mission on STS 62A in 1986.

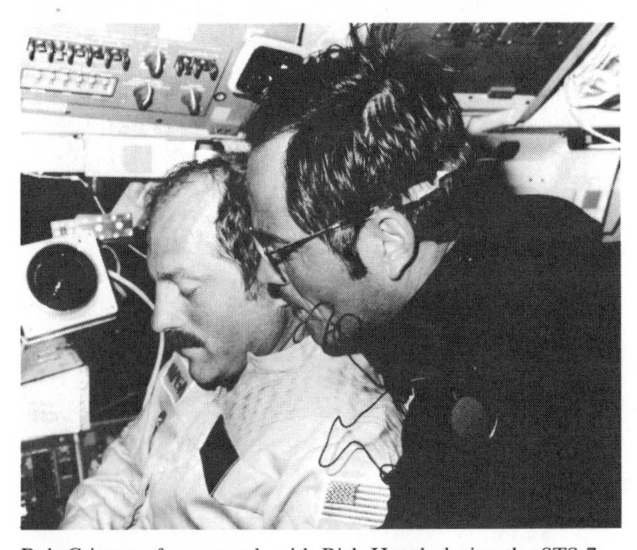

Bob Crippen, foreground, with Rich Hauck during the STS 7 mission. (*NASA*)

The first manned space station was Salyut 1, launched on 19 April 1971 and occupied by the crew of Soyuz 11 on 7 June 1971 for a period of 21 days. This was the only crew to inhabit Salyut 1 and they died on the way home. The first US space station was Skylab 1 launched on 25 May 1973.

The first Soviet-manned spacecraft to splash down was Soyuz 23, launched on 14 October 1976. During an emergency landing, at night, the capsule came down in Lake Tengiz.

Splashdown training for Soyuz crew. (*Novosti*)

The first spaceman with the same surname as a previous spaceman was Vladimir Titov, commander of Soyuz T8, launched on 20 April 1983. He is in fact a relative of Gherman Titov of Vostok 2. The first US spaceman with the same surname was Robert Gibson, pilot of STS 41B, who followed Skylab 4 science pilot Edward Gibson into space.

The first telecast from a manned spacecraft was made by Andrian Nikolyev, the pilot of Vostok 3, launched on 11 August 1962. The first by an American was made by Gordon Cooper, pilot of *Faith 7*, on 15 May 1963 and it was transmitted live to the Earth via Telstar 2.

The first televised lift-off from the Moon was that of Apollo 15 Lunar Module ascent stage, Falcon, in August 1971, taken by a camera mounted on the lunar rover. Similar lift-offs took place on Apollo 16 and 17.

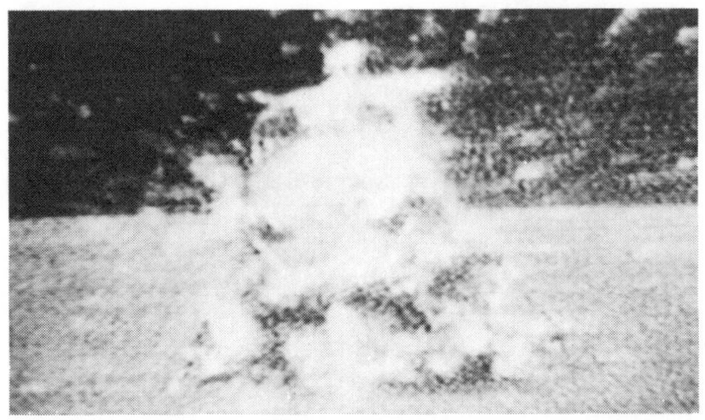

Lift-off of Apollo 15's lunar module ascent stage. (*NASA*)

The first man to make three space flights was Wally Schirra, commander of Apollo 7 on 11 October 1968. His previous flights were on MA 8 in 1962 and Gemini 6, making him the only astronaut to fly all three types of spacecraft. The first Russians to make three flights were Shatalov and Yeliseyev on Soyuz 10 launched on 23 April 1971.

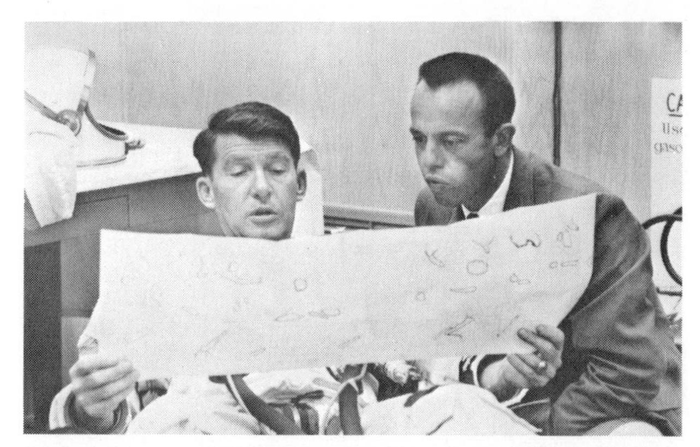

Wally Schirra, left, the first man to fly three missions and Alan Shepard the first to manoeuvre a spacecraft. (*NASA*)

The first flight by three manned spacecraft together took place on 13 October 1969 when Soyuz 8 joined Soyuz 6 and 7 already in orbit. The craft did not dock but Soyuz 7 and 8 did rendezvous at 1600 ft. Seven men were in space for the first time: Shatalov, Yeliseyev, Filipchenko, Gorbatko, Volkov, Shomin and Kubasov.

The first person to land (at sea) in his spacecraft was Alan Shepard, launched on 5 May 1961. The first Russians were the Voskhod 1 crew in 1964. The first Americans to land on land were Young and Crippen at the end of the first shuttle flight in 1981.

The first manned spacecraft to be provided with electrical power by a system other than batteries was Gemini 5, launched on 21 August 1965, and equipped with fuel cells. The first flight by a manned spacecraft equipped with solar cells for power was Soyuz 1 in 1967. During the Apollo 12 mission in 1969, nuclear power was used to operate the instrument-array set down on the lunar surface by the astronauts.

The first three-man space launch was that of Voskhod 1. It carried Vladimir Komarov, Konstantin Feoktistov and Boris Yegerov on 12 October 1964.

The first man to make two flights to the Moon was James Lovell, who was senior pilot of Apollo 8 (launched on 21 December 1968, making ten orbits) and commander of Apollo 13 (launched on 11 April 1970 which, during its emergency return to Earth, flew round the Moon once but did not orbit).

The first man to enter space twice was Gus Grissom, as commander of Gemini 3 on 23 March 1965. He had previously flown *Liberty Bell 7* on a ballistic flight. (The first man in orbit twice was Gemini 5 commander Gordon Cooper.) The first Russian in space twice was Vladimir Komarov of Soyuz 1.

The first flight of over 200 days was made by the Soyuz T5 cosmonauts Anatoli Berezovoi and Valentin Lebedev, who were launched on 13 May 1982.

The first two-man space launch was made by Voskhod 2 on 18 March 1965, crewed by Pavel Belayev and Alexei Leonov.

The first space flight with two women on board was that of STS41G, a *Challenger* shuttle mission launched on 5 October 1984. The two women flew with five men making the largest space crew so far. They were Sally Ride, the first American woman to make a second flight, and Kathy Sullivan who became the first American lady to walk in space during the mission.

The first launch of a manned Soviet spaceflight to be shown on television was that of Soyuz 3 on 26 October 1968.

The televised lift-off of Soyuz 3. (*Novosti*)

The first launch of an American manned spaceflight to be shown 'live' in Europe was that of Gemini 4 on 3 June 1965. The pictures were beamed across the Atlantic by Early Bird. Wally Schirra's launch on MA-8 on 3 October 1962 had been the first to be shown in Europe on television on the day of launch when a recording of the launch was beamed by Telstar a few hours later.

The first man to walk in space was Soviet pilot of Voskhod 2, Alexei Leonov, on 18 March 1965. He spent 10 minutes outside attached to Voskhod by a 15 ft long tether but the total time spent out of the pressurized environment of his cabin was 23 min 41 s. The first American to walk in space was Edward White in June 1965. White used a hand-held manoeuvring unit.

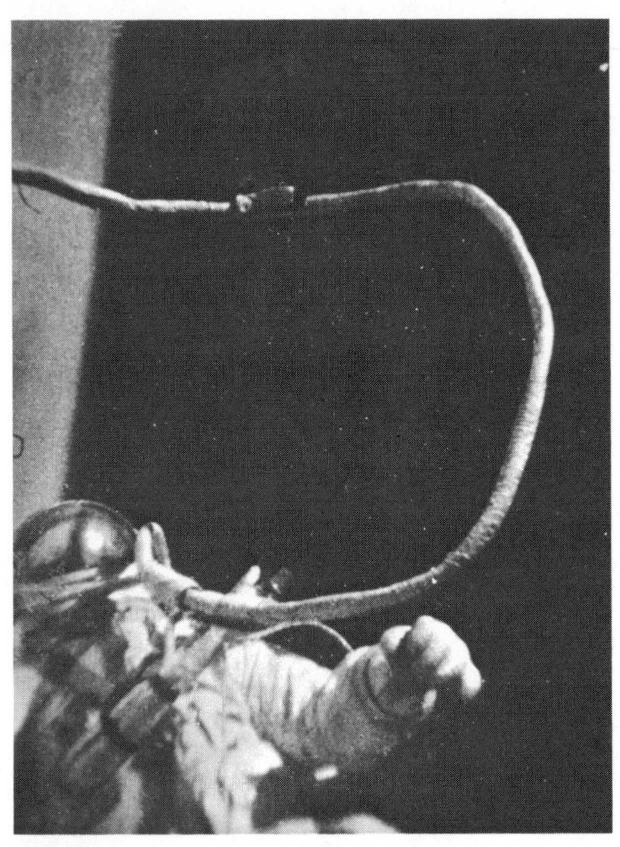

Alexei Leonov – the first spacewalker. (*Novosti*)

The first Western European to enter space was Frenchman Jean Loup Chrétien who flew as research cosmonaut on Intercosmos flight Soyuz T6, on 24 June 1982. The first Western European to be launched by the USA was West German scientist Ulf Merbold, who was the payload specialist of STS 9 – Spacelab 1.

The first woman in space – and the first non-pilot – was Valentina Tereshkova, the pilot of the final Vostok flight, no. 7 on 16 June 1963. The next woman to fly was Svetlana Savitskaya – nearly 20 years later. The first US woman in space was Sally Ride of STS 7 on 18 June 1983.

The first woman to enter space twice was Svetlana Savitskaya, flight engineer of Soyuz T12, launched on 17 July 1984. She was the research engineer on Soyuz T6. The first US woman to make two flights was Sally Ride on STS 7 and 41G.

The first woman to make a spacewalk was Svetlana Savitskaya, during the Soyuz T12 mission to Salyut 7, launched on 17 July 1984. The first US woman to walk in space was Kathryn Sullivan on STS 41G, also in 1984.

A wave to television viewers from Valentina Tereshkova. (*Novosti*)

The first flight in a spacecraft unable to survive re-entry took place during Apollo 9, launched on 3 March 1969, when astronauts McDivitt and Schweickart flew the lunar module in Earth orbit during a test flight which simulated a landing on and take-off from the Moon.

The Apollo 9 lunar module Spider in flight. (*NASA*)

Svetlana Savitskaya outside Salyut 7. (*Novosti*)

The first spaceman to return to Earth unconscious was Vance Brand, Apollo 18 command module pilot, launched on 15 July 1975, who choked on leaking nitrogen tetroxide gas during landing. He and his two colleagues recovered from the after-effects of the gas.

The first flight by a crew not wearing spacesuits and with no means of emergency escape if anything went badly wrong during launch was Voskhod 1 on 12 October 1964. The first by America was STS 5 on 11 November 1982.

The first passenger-observer in space was also the first politician, US Rep Senator Edwin 'Jake' Garn, who was launched on shuttle mission 51D on 12 April 1985.

The first shuttle payload specialist to fly into space twice was Charlie Walker of McDonnell Douglas who flew on 41D in 1984 and was launched on mission 51D on 12 April 1985.

The first spaceflight with more than one 50-year-old person on board was that of STS 51B/Spacelab 3, launched on 29 April 1985 with mission specialists Bill Thornton, 56, and Don Lind, 54 and payload specialist Lodewijk van den Berg, 53.

Voskhod 1 crew, from the front to rear, are Vladimir Komarov, Konstantin Feoktistov and Boris Yegerov. (*Novosti*)

3

Manned Space Machines

Vostok

A 7 ft 5 in diameter spherical capsule housed the cosmonaut who was mounted on an ejection seat for possible emergency escape or for a standard landing procedure whereby he and Vostok parted company at 20 000 ft. The capsule was sparsely equipped and weighed less than Mercury at 1765 lb but remained until re-entry attached to a service module containing a retro rocket which was housed within the second stage rocket. So initially the total weight in orbit was about 10 000 lb. A manned Vostok was flown six times, completing from 1 to 81 orbits. Flight was totally automatic. Retro fire was commanded from the ground.

Vostok flight log (manned flights only)

	Date	Time *days h min*	Wt in orbit *lb*
Vostok 1	12 Apr. 1961	1 48	10 419
Vostok 2	6 Aug. 1961	1 1 18	10 431
Vostok 3	11 Aug. 1962	3 22 22	10 410
Vostok 4	12 Aug. 1962	2 22 57	10 423
Vostok 5	14 June 1963	4 23 6	10 441
Vostok 6	16 June 1963	2 22 50	10 392
Total flight time		15 days 22 h 21 min	

Vostok and its instrument stage. (*Novosti*)

Vostok Booster

The first stage of the Vostok booster was the Soviet ICBM SS-6, Sapwood. The second stage was similar to those that launched the first Luna probes to the Moon. The vehicle was called A-1 or SL3 in the West. The first stage was unusual in that it consisted of a hammer-headed central core with four RD 108 engines and four vernier motors and four tapered strap-on tanks each with four RD107 engines and two verniers. Thrust was 1 111 330 lb. The second stage thrust was 12 350 lb. All engines burned liquid oxygen and kerosene. The A-1 with Vostok was 126 ft long and its base was 34 ft wide. (The first stage is still in use today in the Soyuz launch vehicle.)

Mercury

These spacecraft carried the first American astronauts into space between May 1961 and May 1963. Two astronauts flew sub-orbital missions and four flew into orbit, covering a total of 31 orbits. The Mercury capsule was bell-shaped and 9 ft 5 in high, with a maximum diameter across the base of the heatshield of 6 ft 1 in. Weight at lift-off was about 4260 lb, in orbit 2980 lb and on landing 2420 lb. The attitude of the spacecraft could be changed by the release of short bursts of hydrogen peroxide gas from 18 control thrusters located on the conical and cylindrical portions of the craft. These movements could be controlled by the automatic stabilization and control system ASCS, the rate stabilization and control system RSCS or by manual proportional control; fly-by-wire, a manual-electrical system. Emergency escape from the launch vehicle was provided by a launch escape system rocket on top of Mercury.

Mercury flight log (manned flights only)

		Date	Time				Wt in orbit
			days	*h*	*min*	*s*	*lb*
Freedom 7	MR 3	5 May 61			15	28	2845
Liberty Bell 7	MR 4	21 Jul. 61			15	37	2836
Friendship 7	MA 6	20 Feb. 62		4	55	23	2987
Aurora 7	MA 7	24 May 62		4	56	5	2975
Sigma 7	MA 8	3 Oct. 62		9	13	11	3029
Faith 7	MA 9	15 May 63	1	10	19	49	3033

Total manned flight time 2 days 5 h 55 min 33 s.

Mercury capsule Liberty Bell 7. (*NASA*)

Redstone

The Mercury-Redstone was used for sub-orbital flights. The total vehicle was 83 ft high. The Redstone was an Intermediate Range Ballistic Missile (IRBM) and had one engine, a Rocketdyne A-7, with a thrust of 78 000 lb, burning liquid oxygen, ethyl alcohol and water. Launch was from Pad 5, Cape Canaveral.

Atlas

Mercury orbital flights were boosted by the US ICBM, Atlas D. It had a thrust of 367 000 lb and with Mercury on top was 95.3 ft high. Launches were made from Pad 14 at Cape Canaveral. The Atlas was powered by three engines, burning liquid oxygen and kerosene, two were Rocketdyne LR 89s and these were ejected later in flight and the third engine, a central sustainer, called LR 105, continued the launcher's ascent. The Atlas had two powerful vernier motors which ignited at main engine ignition to stabilize the rocket.

Atlas D launching MA-7. (*NASA*)

Voskhod 1

Voskhod 1 was a Vostok spacecraft equipped to fly three persons. No ejection seats were fitted. No spacesuits were worn by the crew. The fixed seats were fitted in a triangular fashion, with the commander on the right, the scientist on the left and the doctor on top. A soft-landing retro rocket was carried and the craft was equipped with a reserve retro rocket, a plastic-cup-shaped device on top of the capsule, for the de-orbit burn.

Gemini

Ten teams of two astronauts trained themselves and tested equipment and methods for the Apollo lunar landing programme in the Gemini spacecraft. Gemini was a conical and bell-shaped vehicle, like Mercury, but had two sections: a re-entry, crew module with re-entry control and rendezvous and recovery equipment, and an adapter module comprising retro rockets and equipment. The craft weighed about 8000 lb in orbit. It was 18.4 ft long and had a maximum diameter of 10 ft. The flight cabin was 11 ft long with a 90 in diameter heatshield. Systems included 100 per cent oxygen environmental control system, two liquid oxygen and liquid hydrogen fuel cells, not carried on Gemini 3 or 4, to generate electricity. Sixteen RCS thrusters were used for manoeuvring and four solid propellant 2500 lb thrust rockets were for the retrofire. Gemini also carried a drogue and main chute which, when opened, caused Gemini to land on the horizontal rather than base-down. Emergency escape was by the use of ejection seat.

Cooper, left, and Conrad inside Gemini 5 before launch. (*NASA*)

Gemini flight log (manned flights only)

	Date	Time				Wt. in orbit
		days	*h*	*m*	*s*	
Gemini 3	23 Mar. 1965		4	52	51	7111
Gemini 4	3 June 1965	4	1	56	12	7879
Gemini 5	21 Aug. 1965	7	22	55	14	7949
Gemini 7	4 Dec. 1965	13	18	35	1	8069
Gemini 6	15 Dec. 1965	1	1	51	54	7817
Gemini 8	16 Mar. 1966		10	41	26	8351
Gemini 9	3 June 1966	3	0	20	50	8087
Gemini 10	18 July 1966	2	22	46	39	8295
Gemini 11	12 Sep. 1966	2	23	17	8	8374
Gemini 12	11 Nov. 1966	3	22	34	31	8297

Total manned flight time 40 days 9 h 51 min 46 s.

Gemini 6 and 7 during first space rendezvous. (*NASA*)

A Voskhod mock-up. The cosmonaut in the foreground is Boris Volynov who was back-up commander of Voskhod 1 and the commander of the cancelled flight of Voskhod 3. Note the similarly dressed cosmonaut (Shonin?) at the rear. (*Novosti*)

Voskhod 2

This spacecraft was equipped with two seats instead of three. Room for the third seat was taken by the exit hatch for a telescopic inflatable airlock approximately 7 ft long and 3 ft wide. A television camera was mounted on the reserve retro pack and a cine camera at the end of the airlock, which was jettisoned in flight after the EVA. There was no provision for emergency ejection.

Voskhod flight log (manned flights only)

	Date	Time *days h mins s*	Wt in orbit *lb*
Voskhod 1	12 Oct. 1964	1 0 17 3	11 731
Voskhod 2	18 Mar. 1965	1 2 2 17	12 529

Total manned flight time 2 days 2 h 19 min 20 s.

The launching of Voskhod 2. (*Novosti*)

A2

Voskhod, Soyuz and Progress vehicles are launched by this uprated Vostok booster called the A2–SL4. It has the same first stage but a more powerful second stage with a thrust of 66 150 lb. With Voskhod the vehicle was 143 ft high and with Soyuz 162 ft high.

The A2 launches Soyuz 35. (*Novosti*)

Titan 2

Gemini was launched by the Titan ICBM. This was 109 ft tall and 10 ft in diameter. It had two first-stage engines called LR-87s with a total thrust of 430 000 lb and a second stage engine, LR-91, with a thrust of 100 000 lb. Both stages were powered by nitrogen tetroxide and hydrazine and a mixture of undemethyl hydrazine, hypergolic chemicals that ignited spontaneously on contact with a characteristic high-pitched whine.

Titan 2 launches Gemini 11. (*NASA*)

Agena

The Lockheed Gemini Agena Target Vehicle, GATV, was an Agena D stage (launched by Atlas) equipped with a docking collar and rendezvous electronics. It was 36 ft 3 in long and 5 ft in diameter and weighed 7000 lb. One Bell 8096 rocket engine had a thrust of 16 000 lb firing on inhibited red fuming nitric acid (IRFNA) and unsymmetrical dimethyl hydrazine (UDMH). There was a secondary propulsion system comprising two 200 lb and two 16 lb thrust motors burning mixed oxides of nitrogen (MON) and UDMH. All motors were restartable. Total length of the Atlas Agena D vehicle was 104 ft.

Gemini Agena Target Vehicle flight log

	Date	
Agena 6	25 Oct. 1965	Destroyed during launch
Agena 8	16 Mar. 1966	Docked GT8, Rendezvous GT10 Re-entered 15 Sep. 67
Agena 9	17 May 1966	Destroyed during launch
Agena 10	18 July 1966	Docked GT10, Restart Re-entered 29 Dec. 66
Agena 11	12 Sep. 1966	Docked GT11, Restart Re-entered 30 Dec. 66
Agena 12	11 Nov. 1966	Docked GT12 Re-entered 23 Dec. 66

Agena 12. (*NASA*)

ATDA

Built to act as a reserve target vehicle for Gemini, ATDA (Augmented Target Docking Adapter) was an unpowered vehicle launched by Atlas. It was 10 ft 11 in long, with a diameter of 5 ft. ATDA weighed 1700 lb in orbit. The total Atlas ATDA vehicle was 95 ft 6 in high. Used only once, ATDA failed to shed its payload shrouds preventing Gemini 9 from docking with it.

ATDA log

ATDA1/GT9	1 June 1966	Rendezvous GT9	Re-entered 11 June 66

The ATDA in space. (*NASA*)

Soyuz (Independent)

Soyuz was 26 ft long and weighed 14 000 lb. It consisted of three modules and derived its electrical power from two solar panels. At the front of the craft was an orbital module which had at its front a docking probe for insertion into another Soyuz (which would be fitted with the female attachment) or Salyut. The orbital module acted as an airlock on joint flights and as a laboratory on solo missions. Attached to this spherical module was a descent flight crew module which was shaped like an inverted cup. This was equipped with two parachutes, one an emergency, and a drogue which opened at 25 000 ft. A retro rocket fired just before landing about 3 ft above the ground to reduce the impact speed to 1 ft/sec. Behind this was an instrument section to which were attached two 12 ft long foldable solar panels. The instrument section was equipped with two 880 lb rocket motors, one a back-up. Atop an A2 booster the Soyuz was encased in a payload shroud at the top of which was a three tier rocket escape tower. (Basic changes were made to the system through 1–22.)

A Soyuz being assembled. Note solar panels. (*Novosti*)

Soyuz 9 descent capsule. (*Novosti*)

Soyuz (independent) manned flight log

	Date	days	h	min	s	
Soyuz 1	23 Apr. 67	1	2	45		Crashed
Soyuz 2A						Flight cancelled. Was to dock with Soyuz 1
(Soyuz 2)						(Unmanned target for Soyuz 3)
Soyuz 3	26 Oct. 68	3	22	51		Solo
Soyuz 4	14 Jan. 69	2	23	21		Target for Soyuz 5
Soyuz 5	15 Jan. 69	3	0	46		Docked Soyuz 4
Soyuz 6	11 Oct. 69	4	22	42		Joint flight
Soyuz 7	12 Oct. 69	4	22	41		Joint flight
Soyuz 8	13 Oct. 69	4	22	51		Joint flight
Soyuz 9	1 Jun. 70	17	16	58	50	Solo
Soyuz 10	23 Apr. 71	1	23	45		Docked Salyut 1
Soyuz 11*	6 Jun. 71	23	18	22		Docked Salyut 1
Soyuz 13	18 Dec. 73	7	20	55		Solo
Soyuz 15†	2 Dec. 74	5	22	24		ASTP rehearsal
Soyuz 19†	15 July 75	5	22	30	54	ASTP
Soyuz 22	15 Sep. 76	7	21	54		Solo

Total manned mission flight time 97 days 0 h 45 min 44 s.

* Includes time in Salyut.
† Carried special ASTP docking system.

Soyuz 19 in space. (*NASA*)

Apollo Command and Service Module (CSM)

There were 11 manned flights in the Apollo programme, two in Earth orbit and nine to the Moon. Apollo was also used as the Skylab space station ferry and during the ASTP mission. Astronauts flew in the command module (CM) which was a cone 12 ft high and 12 ft 10 in wide. It was equipped with a heatshield, 12 reaction control motors and landing systems and weighed about 12 250 lb. The service module (SM) was 12 ft 10 in wide and 24 ft 7 in high. It weighed about 51 240 lb. The SM incorporated fuel cells for electrical power, a service propulsion system engine of 20 500 lb thrust for lunar orbit insertion, de-orbit and midcourse corrections, and RCS thrusters. Modules on Apollo 15, 16, and 17 included SIM bays, packages of lunar orbit science equipment. Emergency escape was provided by a launch escape rocket on top of the CM.

Apollo 14 in lunar orbit. (*NASA*)

Apollo manned flight log

Designation	Name	Date	Time days	h	min	s	Type of flight	Weight
Apollo 7		11 Oct. 68	10	20	9	3	EO	32 395 lb
Apollo 8		21 Dec. 68	6	3	0	42	LO	63 717 lb
Apollo 9	*Gumdrop*	3 Mar. 69	10	1	0	54	EO	80 599 lb*
Apollo 10	*Charlie Brown*	18 May 69	8	0	3	23	LO	94 512 lb*
Apollo 11	*Columbia*	16 Jul. 69	8	3	18	35	LO	96 715 lb*
Apollo 12	*Yankee Clipper*	14 Nov. 69	10	4	36	25	LO	96 812 lb*
Apollo 13	*Odyssey*	11 Apr. 70	5	22	54	41	LFB	96 964 lb*
Apollo 14	*Kitty Hawk*	31 Jan. 71	9	0	1	57	LO	98 137 lb*
Apollo 15	*Endeavour*	26 Jul. 71	12	7	11	53	LO	103 143 lb*
Apollo 16	*Casper*	16 Apr. 72	11	1	51	5	LO	103 165 lb*
Apollo 17	*America*	7 Dec. 72	12	13	51	59	LO	103 187 lb*
Skylab 2†		25 May 73	28	0	49	49	EO/D	30 384 lb
Skylab 3†		28 Jul. 73	59	11	9	4	EO/D	30 561 lb
Skylab 4†		16 Nov. 73	84	1	15	31	EO/D	30 100 lb
Apollo 18	ASTP	15 Jul. 75	9	1	28	0	EO/D	32 563 lb×

TOTALS

	days	h	min	s
Apollo	104	6	0	37
Skylab	171	13	14	24
ASTP	9	1	28	0

Abbreviations

EO	Earth orbit
LO	Lunar orbit
LFB	Lunar fly by
EO/D	Earth orbit/docking

* With Lunar Module
† Includes time with Skylab
× With docking module.

The Apollo 15 command module. (*NASA*)

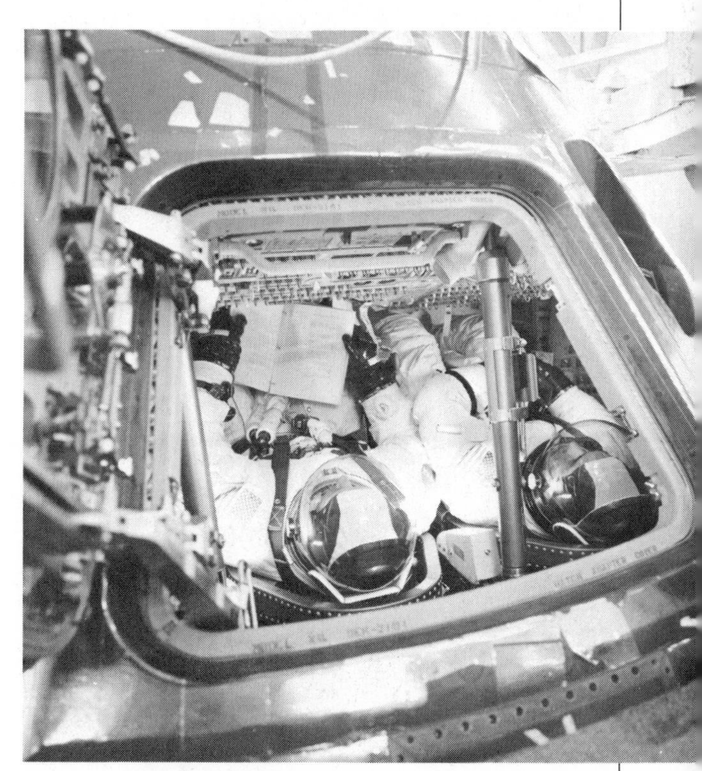
A closer look inside Apollo 15. (*NASA*)

Saturn 1B

This was the rocket that launched five teams of US astronauts – Apollo 7, Skylabs 2, 3 and 4 and Apollo 18 – ASTP. The first stage of the Saturn 1B was 80.3 ft high and 21.4 ft in diameter and was powered by eight H-1 engines developing a thrust of 1 640 000 lb. Propellants were liquid oxygen and RPI. The second stage, the SIVB, was to become the third stage of the Saturn 5 vehicle. This was 59.1 ft long and also 21.4 ft in diameter. Its restartable J2 engine had a thrust of 225 000 lb and used liquid oxygen and liquid hydrogen propellants. Between the first and second stages was an Instrument Unit which controlled the guidance, telemetry, radio frequency, environmental control, electrical power and structural stability of the rocket and spacecraft. With Apollo on top, the total length was 224 ft. It was launched from Pad 34 (Apollo 7) and Pad 39B.

Saturn 1B launches Skylab 2 crew. (*NASA*)

Saturn 5

The mighty Saturn 5 flew 13 times, comprising two unmanned Apollo tests, one to carry Skylab and ten to carry the Apollo astronauts on nine missions to the Moon and one into Earth orbit. With Apollo at its top, Saturn 5 was 363 ft high. The first stage S1C was 138 ft long and 33 ft in diameter. With five F-1 engines (LOX/RP1) at full thrust, it developed 7 650 000 lb. The second stage was SII, 81 ft long and also 33 ft in diameter. Five J2 (LOX/LH) engines gave it a thrust of 1 150 000 lb. The third stage was a 238 000 lb thrust S4B (see Saturn 1B). With the Instrument Unit and complete Apollo combination, the huge vehicle weighed about 6 400 000 lb. On Skylab the rocket was 333 ft high and weighed 6 222 000 lb.

S4B stages on manned flights

Apollo 8	Solar orbit
Apollo 9	Solar orbit
Apollo 10	Solar orbit
Apollo 11	Solar orbit
Apollo 12	Solar orbit
Apollo 13	Lunar impact – 15 Nov. 70
Apollo 14	Lunar impact – 04 Nov. 71
Apollo 15	Lunar impact – 30 Jul. 71
Apollo 16	Lunar impact – 19 Apr. 72
Apollo 17	Lunar impact – 10 Dec. 72

The S4B stage from Skylab 3's Saturn 1B. (*NASA*)

Saturn 5 launches Apollo 15. (*NASA*)

Apollo 11's lunar module Eagle with Buzz Aldrin in foreground. (*NASA*)

Apollo Lunar Module

The Apollo Lunar Module or LM was 22 ft 11 in high and consisted of two major components, the descent stage and ascent stage. The descent stage was 10 ft 7 in high and 14 ft 1 in wide; its four landing legs had footpads. The descent engine had a thrust of 10 000 lb. In the side of this stage much scientific equipment was stowed, including an ingeniously packed foldable lunar roving vehicle on Apollo 15, 16 and 17. The ascent stage used the descent stage as a launching pad and flew skywards beneath a thrust of 3500 lb. It was 12 ft 4 in high and 14 ft 1 in wide. The LM was housed beneath the CM/SM for launch and was pulled out of the S4B during a transposition and docking manoeuvre.

Manned flight log – Apollo Lunar Module (Independent manned flight time, inc. on lunar surface)

				Time		
				days	*h*	*min*
Apollo 9	*Spider*	EO	3 Mar. 69		6	20
Apollo 10	*Snoopy*	LO	18 May 69		8	
Apollo 11	*Eagle*	LL	16 July 69	1	3	51
Apollo 12	*Intrepid*	LL	14 Nov. 69	1	13	42
Apollo 13	*Aquarius*	LFB*	11 Apr. 70			
Apollo 14	*Antares*	LL	31 Jan. 71	1	15	45
Apollo 15	*Falcon*	LL	26 July 71	3	0	56
Apollo 16	*Orion*	LL	16 Apr. 72	3	9	28
Apollo 17	*Challenger*	LL	7 Dec. 72	3	8	10
Total				14 days 18 h 12 min		

* Not manually flown solo.

Descent stage

Spider	Decayed 18 March 1969
Snoopy	In lunar orbit
Eagle	Tranquility Base 0° 41' N 23° 36' E
Intrepid	Ocean of Storms 3° 11' S 23° 23' W
Aquarius	Decayed 15 April 1971
Antares	Fra Mauro 3° 40' S 17° 28' W
Falcon	Hadley Base 26° 5' N 3° 40' E
Orion	Descartes 8° 59' S 15° 30' E
Challenger	Taurus Littrow 20° 10' N 30° 45' E

Ascent stage

Spider	Decayed 22 March 1969
Snoopy	Solar orbit
Eagle	Lunar orbit

Intrepid	Lunar impact – 20 November 1969
Aquarius	Decayed – 15 April 1971
Antares	Lunar impact – 7 February 1971
Falcon	Lunar impact – 3 August 1971
Orion	Lunar impact – 29 May 1972
Challenger	Lunar impact – 15 December 1972

The Apollo 16 lunar module ascent stage. (*NASA*)

MET

The MET (Modularized Equipment Transporter) was a two-wheeled handcart used by the Apollo 14 astronauts to carry tools and samples around during their two walks on the Moon in February 1971. It was left on the Moon at Fra Mauro.

Salyut 1

Salyut was the first space station. With a Soyuz craft attached to one end it weighed 55 125 lb and was 65 ft long. The space station itself was 47.3 ft long and comprised of various sections: docking port, transfer compartment, work station and an instrument section. At its narrowest it measured 6.56 ft in diameter and at its widest 13.6 ft. Power was provided by two pairs of solar panels, one at the front and one at the rear. Salyut had one docking port at its front and at its rear a manoeuvring engine capable of orbital height alterations.

Log

Launched:	19 April 1971
Re-entered:	11 October 1971

(Deliberate de-orbit and re-entry into unpopulated area)

Visitors

Soyuz 10	(Crew docked but did not transfer)
Soyuz 11	

Soyuz 11 prepares to dock with Salyut 1 in this artist's impression. (*Novosti*)

Lunar Rover

The lunar roving vehicle was ingeniously packed into the side of the lunar module so that with a pull of a few lanyards it was deployed on the surface. It was 10 ft long and 6 ft 9 in wide and had four wire mesh wheels. Powered by two silver zinc batteries producing 36 volts, it was able to travel at a maximum speed of 8.7 mph. Unloaded, the lunar rover weighed 460 lb but had a capacity to hold over 1000 lb of cargo. A large dish aerial provided direct communications with Earth and an on-board television camera transmitted the first pictures of a lunar take-off.

Log

Apollo LRV 15	Hadley Rille	Parked at 26° 5' N 3° 40' E
Apollo LRV 16	Descartes	Parked at 8° 59' S 15° 30' E
Apollo LRV 17	Taurus Littrow	Parked at 20° 10' N 30° 45' E

The Apollo 15 Lunar Roving Vehicle. (*NASA*)

Apollo Sub-satellite

On the flights of Apollo 15 and 16, a sub-satellite was sprung from the service module SIM bay in lunar orbit. Weighing about 80 lb, the satellite was used to measure areas of the Moon called mascons which exerted a more powerful gravitational pull. The satellite was spin stabilized on deployment and was hexagonal in shape and had three booms on which one had a magnetometer.

Flight log

Apollo 15	Lunar orbit
Apollo 16	Lunar impact

Soyuz–Salyut Ferry Vehicle

This Soyuz craft weighing about 15 000 lb carried no solar panels, relying instead on internal chemical batteries which could be recharged from the solar power supply on Salyut. But because of this, the craft was limited to two days' independent flight, by which time it had to have made a docking. If not, the Soyuz came home immediately.

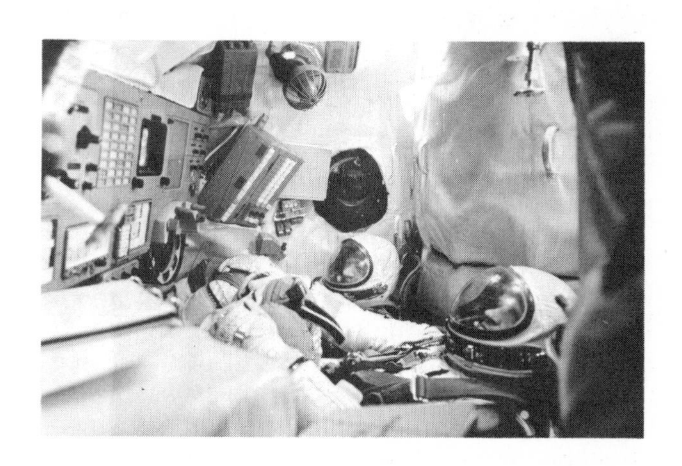

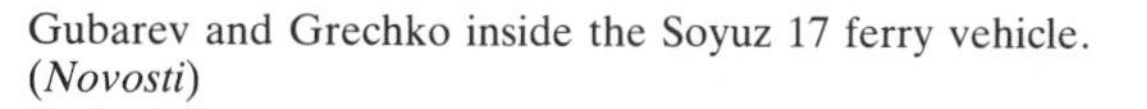

Gubarev and Grechko inside the Soyuz 17 ferry vehicle. (*Novosti*)

Manned flight log

Craft	Date	Mission time				Actual time*	Mission
		days	*h*	*min*	*s*	*days*	
Soyuz 12	27 Sep. 1973	1	23	16		—	Solo test
Soyuz 14	3 July 1974	15	17	30		—	Salyut 3
Soyuz 15	26 Aug. 1974	2	0	12		—	Failed to dock
Soyuz 17	11 Jan. 1975	29	13	20		—	Salyut 4
Soyuz 18-1	5 Apr. 1975			21	27	—	Launch abort
Soyuz 18	24 May 1975	62	23	20		—	Salyut 4
(Soyuz 20)							(Unmanned test of Progress tanker)
Soyuz 21	6 July 1976	49	6	24		—	Salyut 5
Soyuz 23	14 Oct. 1976	2	0	6		—	Failed to dock
Soyuz 24	7 Feb. 1977	17	17	26		—	Salyut 5
Soyuz 25	9 Oct. 1977	2	0	46		—	Failed to dock Salyut 6
Soyuz 26	10 Dec. 1977	96	10			37	Salyut 6*
Soyuz 27	10 Jan. 1978	5	22	59		65	Salyut 6*
Soyuz 28	2 Mar. 1978	7	20	16		—	Salyut 6
Soyuz 29	15 June 1978	139	14	48		80	Salyut 6*
Soyuz 30	27 June 1978	7	22	4		—	Salyut 6

Skylab

Born as the Apollo Applications Programme to utilize many leftover parts of the Moon landing project, the Skylab space station was essentially a converted S4B stage, with additional modules attached, one of which utilized LM hardware. It was launched by Saturn 5. Astronauts flew to it in Apollo command and service modules launched by Saturn 1Bs. When the CSM was attached to Skylab its length was 118.5 ft, with a working volume of about 13 000 cu ft. It was also the second heaviest object in space at about 200 000 lb with the CSM attached. The first module was the MDA (Multiple Docking Adapter) with two docking ports which was 17.3 ft long and 10 ft in diameter. Attached to this, mounted on a truss framework, was the Apollo Telescope Mount which was not a habitable module but a solar observatory. This was about 13 ft long and had four extendable solar arrays. Attached longitudinally to the MDA was an Airlock Module which was 17.6 ft long and 10 ft in diameter. Then came the Instrument Unit, 3 ft long and 21.6 ft in diameter. Finally the Orbital Workshop itself, 48.1 ft long and with a diameter of 21.6 ft. Over this was mounted a meteroid-thermal shield and from it extended two solar panels. However, the shield and one panel were torn loose during launch and one panel failed to deploy in orbit. The first astronaut crew erected a temporary shield and deployed the remaining solar panel.

Skylab 1 launched 14 May 1973 – decayed naturally 11 July 1979

Manned by teams from	**Time** (*dock to undock*) *days*	*h*	*min*	*s*
Skylab 2	27	6	48	7
Skylab 3	59	0	9	42
Skylab 4	83	12	32	12

513 man-days in space (see Apollo CSM)

Skylab from the mission 4 crew. (*NASA*)

Skylab from mission 3 crew. (*NASA*)

Soyuz 31	26 Aug. 1978	7	20	49	68	Salyut 6*
Soyuz 32	25 Feb. 1979	175	0	36	108	Salyut 6*
Soyuz 33	10 Apr. 1979	1	23	1	—	Failed to dock
(Soyuz 34)						(Unmanned. Brought home S32 crew)
Soyuz 35	9 Apr. 1980	184	20	12	55	Salyut 6*
Soyuz 36	26 May 1980	7	20	46	66	Salyut 6*
Soyuz 37	23 July 1980	7	20	42	80	Salyut 6*
Soyuz 38	18 Sep. 1980	7	20	43	—	Salyut 6
Soyuz 39	22 Mar. 1981	7	20	43	—	Salyut 6
Soyuz 40	15 May 1981	7	20	28	—	Salyut 6
Total flight time		849 days 20 h 48 min 27 s				

* Crews swopped craft for re-entry.

Salyut 3

The second Salyut failed in orbit. Salyut 3, its replacement, was a military space base and differed somewhat from Salyut 1. Instead of four solar panels, two at the front and two at the rear, Salyut 3 was equipped with one pair at the rear of the station and comprised two cylindrical compartments. Salyut 3 was equipped with a docking port at the rear so that the front was freed for returnable film pods.

Launched 25 June 1974
Re-entered (by command) 24 January 1975

Visiting craft

Soyuz 14
Soyuz 15 (failed to dock)
Military film pod ejected September 1974

Salyut 4

Salyut 4 was equipped for conventional docking at the front and was used for civilian missions, rather than military ones. Three solar panels were mounted in the centre of the three-compartment station, and this could automatically rotate at 180° to catch the sun.

Launched 26 December 1974
Re-entered (by command) 3 February 1977

Visiting craft

Soyuz 17
Soyuz 18-1 Launch abort
Soyuz 18
Soyuz 20 Unmanned test of Progress tanker

Salyut 4. (*Novosti*)

ASTP Docking Module

Flown just once on the joint Apollo–Soyuz mission, the docking module (DM) flew with Apollo in the same way the lunar module did. Housed beneath the CSM during launch, it was plucked out of the Saturn 1B's S4B stage during a transposition and docking manoeuvre. The Apollo end of the module used a standard LM-style docking facility (actually flown on Apollo 14) and at the other, a docking system compatible with Soyuz's special arrangement. The DM was 10.3 ft long and 4.7 ft in diameter. It weighed 7390 lb.

Launched	15 July 1975
Re-entered	2 August 1975

ASTP docking system together with Soyuz 4 descent capsule at Star City. (*Author*)

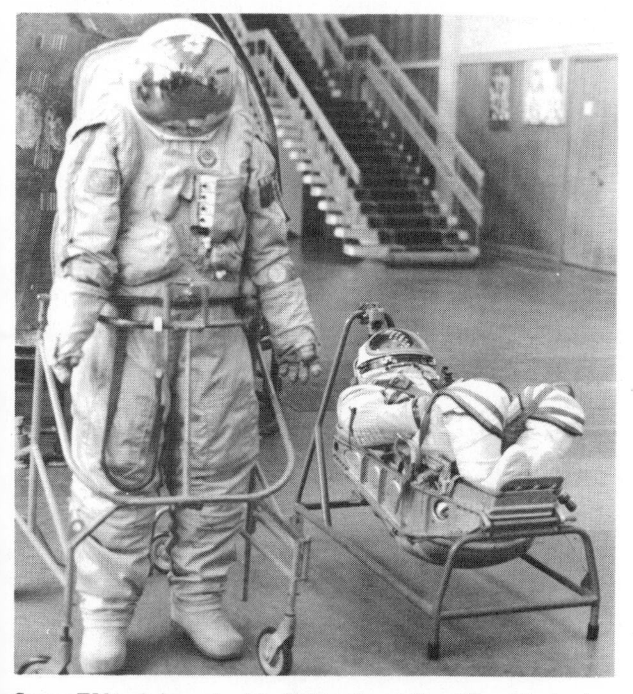

Soyuz EVA, left, and cabin flight suits at Star City. (*Author*)

Progress

These unmanned tanker craft were vital to the long-duration crews aboard Salyut space stations. They are essentially Soyuz ferry vehicles but with no need for an emergency escape rocket on the launch vehicle or for a re-entry heatshield. More cargo weight is therefore available. Progress is a little longer than standard Soyuz. It has an independent flying time of 8 days and can remain attached to Salyut for months. It carries 2600 lb of dry cargo plus 2025 lb of fuel, for the space station's propulsion systems, or water. Progress demonstrated the first refuelling of craft in space. It is 26 ft long with a maximum diameter of 8.92 ft and weighs 15 479 lb at launch. Thirty per cent of weight is cargo.

Progress tanker flight log

	Launch	Lifetime (days)	Salyut
Soyuz 20	17 Nov. 1975	90	4
Progress 1	20 Jan. 1978	18	6
2	7 July 1978	28	6
3	8 Aug. 1978	16	6
4	4 Oct. 1978	22	6
5	12 Mar. 1979	24	6
6	13 May 1979	27	6
7	28 June 1979	20	6
8	27 Mar. 1980	30	6
9	27 Apr. 1980	25	6
10	19 June 1980	20	6
11	29 Sep. 1980	73	6
12	24 Jan. 1981	52	6
13	23 May 1982	14	7
14	10 July 1982	32	7
15	18 Sep. 1982	28	7
16	31 Oct. 1982	43	7
17	17 Aug. 1983	32	7
18	20 Oct. 1983	27	7
19	21 Feb. 1984	41	7
20	15 Apr. 1984	22	7
21	8 May 1984	16	7
22	28 May 1984	?	7

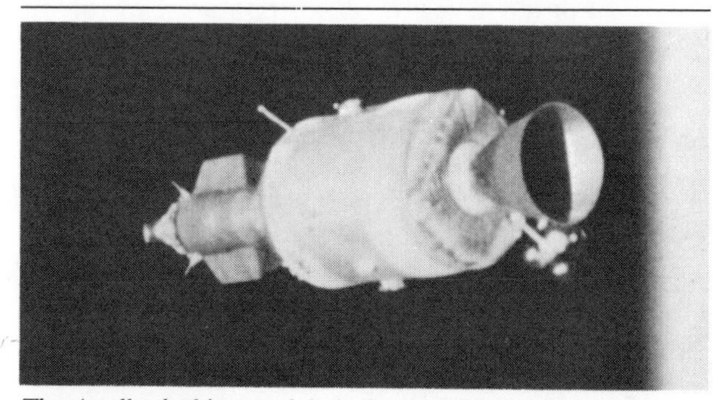

The Apollo docking module in front of CSM. (*NASA from USSR crew photo*)

Salyut 5

See Salyut 3 Military space base
Launched 22 June 1976
Re-entered (by command) 8 August 1977

Visiting craft

Soyuz 21
Soyuz 23 Failed to dock
Soyuz 25
Military film pod ejected from Salyut 5 in March 1977

Salyut 6

The major difference of this space station from its predecessors was its secondary docking port at the rear, facilitated by the removal of Salyut's centrally mounted main engine. This was replaced by two main propulsion engines on either side of this docking port. Salyut's length was 49.2 ft with a minimum diameter of 6.56 ft and maximum 13.6 ft. It weighed 41 674 lb. It was also equipped with many EVA handrails.

Launched 29 September 1977
Re-entered (using Star module retrofire) 28 July 1982

Visitors to Salyut 6

Soyuz 25	Failed to dock
Soyuz 26	
Soyuz 27	
Progress 1	
Soyuz 28	
Soyuz 29	
Soyuz 30	
Progress 2	
Progress 3	
Soyuz 31	
Progress 4	
Soyuz 32	
Progress 5	
Soyuz 33	Failed to dock
Progress 6	
Soyuz 34	Unmanned
Progress 7	
Soyuz T-1	Unmanned
Progress 8	
Soyuz 35	
Progress 9	
Soyuz 36	
Soyuz T-2	
Soyuz 37	
Soyuz 38	
Progress 11	
Progress 12	
Soyuz T-3	
Soyuz T-4	
Soyuz 39	
Soyuz 40	
Cosmos 1267	Unmanned Star module

Salyut 6 was occupied for a total of 676 days and was home to 16 cosmonaut crews.

Soyuz T

In terms of shape, volume and weight, Soyuz T is the same as the Soyuz ferry vehicle but with three important differences – it can carry a crew of three, can dock automatically with Salyut and carries solar panels. It also has an engine system compatible with the Salyut propulsion system. Also, the orbital module is discarded prior to retrofire, and not afterwards. This means that Soyuz T has the facility to leave orbital modules attached to Salyut stations.

Log

	Date	Time			Actual time	
		days	*h*	*min*	*days*	
Soyuz T 1	(Unmanned)	—	—	—	—	Salyut 6
Soyuz T 2	5 June 1980	3	22	19	—	Salyut 6 (2 crew)
Soyuz T 3	27 Nov. 1980	12	19	8	—	Salyut 6
Soyuz T 4	12 Mar. 1981	74	17	38	—	Salyut 6 (2 crew)

Above Fra Mauro Base, 6 February 1971. Take-off. Lunar Module Antares takes off after a stay of 33 h on the surface of the Moon. This was the third successful landing on the lunar surface. Already, public and political interest was waning and later Apollo flights had been cancelled. (*NASA*)

Kennedy Space Centre, 26 July 1971. Rovers to the Moon. Despite sagging interest, thousands of people still flocked to the Kennedy Space Centre to witness the spectacular launches of Saturn 5. This photo was taken from the press site, about 2¾ miles from Pad 39A. Apollo landed at Hadley Base and the two explorers drove about the Moon in the first lunar roving vehicle. (*Author*)

Over the Moon, December 1972. Earthrise. One of the most spectacular sights of the Apollo programme. (*NASA*)

Taurus Littrow Base, December 1972. Apollo 17. The final flight to the Moon, probably this century. The days of space exploration were ending, the days of exploitation just beginning. (*NASA*)

	Date	Time days h min	Actual time days	
Soyuz T 5	13 May 1982	211 8 5	106	Salyut 7 (2 crew)*
Soyuz T 6	24 June 1982	7 22 42	—	Salyut 7
Soyuz T 7	19 Aug. 1982	7 21 52	113	Salyut 7*
Soyuz T 8	20 Apr. 1983	2 0 20	—	Failed to dock Salyut 7
Soyuz T 9	27 June 1983	149 10 46	—	Salyut 7 (2 crew)
Soyuz T 10-1	27 Sep. 1983	Launch pad abort	—	Salyut 7
Soyuz T 10	8 Feb. 1984	236 22 50	—	Salyut 7
Soyuz T 11	3 Apr. 1984	7 21 41	—	Salyut 7
Soyuz T 12	17 July 1984	11 19 14	—	Salyut 7

*Swopped descent craft.

Total manned flight time 726 days 18 h 35 min

Star Module

Space station tug and electrical power module, 40 ft long, Star carries a bell-shaped re-entry vehicle capable of returning 1100 lb of cargo to Earth. It produced 3 kw of electricity from 40 sq metres of solar cells. It can carry 2½ times the amount of cargo on Progress. Star measured 42.6 ft by 13.1 ft, adding 50 cu metres work area to the station.

Cosmos 1267	Launched 25 April 1981 Docked 19 June 1981 Re-entered with Salyut 6 28 July 1982
Cosmos 1443	Launched 2 March 1983 Docked 10 March 1983 Salyut 7 Re-entered 19 September 1983 Ejected descent module 23 August 1983

Space Shuttle

A partly reusable space transportation system, comprising an Orbiter spaceplane, capable of flying 65 000 lb of payload into space and returning with 32 000 lb, an external fuel tank called ET and two solid rocket boosters, SRBs. The orbiter weighs about 250 000 lb. The whole system 4.5 million lb.

The orbiter is 122.2 ft long and has a wing span of 78.06 ft. The SRBs are 149.17 ft long and 12.17 ft in diameter. The ET is 154.2 ft tall and 27.5 ft wide. Total length of the system is 184.2 ft.

The SRBs weigh 1 259 000 lb, have a thrust of 3 300 000 lb each burning ammonium perchlorate oxidizer, aluminium fuel, iron oxide catalyst, a polymer binder and an epoxy curing agent. They burn for 127 s and are then ejected. They are recoverable and some parts reusable. The expendable ET weighs 1 655 600 lb and contains 1 361 936 lb of liquid oxygen and 227 641 lb of liquid hydrogen. This is fed into the orbiter's three space shuttle engines, SSMEs which are throttleable from 65 per cent to 109 per cent power, generating a maximum thrust of 375 000 lb.

The orbiter has two orbital manoeuvring system engines (OMS) which make the final burn to orbit and retrofire. Orientation manoeuvres are achieved by 38 primary and 6 vernier thrusters of Reactor Control System. What makes the orbiter reusable – apart from the reusable main engine – is its Thermal Protection System of ceramic-fibre and carbon-carbon tiles and thermal blankets, which protect the orbiter from temperatures varying from 700 °F and 2300 °F during re-entry.

Shuttle is launched from Pad 39A and B Kennedy Space Centre and SLC 6 'Slick 6' at Vandenberg AFB, California, although 39B and SLC 6 launches have not taken place yet.

Orbiter fleet

Enterprise (used for ALT) and space vehicles *Columbia, Challenger, Discovery* and *Atlantis* (not yet flown).

Space Shuttle flight log

	Date	Time days	h	min	s	Landings*	
STS 1	12 Apr. 1981	2	6	20	52	(E)	*Columbia 1*
STS 2	12 Nov. 1981	2	6	13	11	(E)	*Columbia 2*
STS 3	22 Mar. 1982	8	0	4	46	(WS)	*Columbia 3*
STS 4	27 June 1982	7	1	9	31	(E)	*Columbia 4*
STS 5	11 Nov. 1982	5	2	14	26	(E)	*Columbia 5*
STS 6	4 Apr. 1983	5	0	23	42	(E)	*Challenger 1*
STS 7	18 June 1983	6	2	24	10	(E)	*Challenger 2*
STS 8	30 Aug. 1983	6	1	8	40	(E)	*Challenger 3*
STS 9	28 Nov. 1983	10	7	47	23	(E)	*Columbia 6*
STS 41B	3 Feb. 1984	7	23	15	54	(K)	*Challenger 4*
STS 41C	6 Apr. 1984	6	23	40	5	(E)	*Challenger 5*
STS 41D	30 Aug. 1984	6	0	56	4	(E)	*Discovery 1*
STS 41G	5 Oct. 1984	8	5	23	33	(K)	*Challenger 6*
STS 51A	8 Nov. 1984	7	23	45	54	(K)	*Discovery 2*
STS 51C	24 Jan. 1985	3	1	33	13	(K)	*Discovery 3*
STS 51D	12 April 1985	6	23	56		(K)	*Discovery 4*
STS 51B	29 April 1985	7	0	8	50	(E)	*Challenger 7*

Total manned flight times

	days	h	min	s
Columbia	34	23	50	9
Challenger	47	8	24	54
Discovery	24	2	31	11
Total	106	10	46	14

*Landings

E Edwards Air Force Base
WS White Sands
K Kennedy Space Centre

Inside Salyut 7 mock-up are Soyuz T11 crew, left to right, Gennadi Strekalov, Rakesh Sharma, from India and Yuri Malyshev. (*Novosti*)

A mock-up of Salyut 7 at Star City. (*Author*)

Salyut 7

Similar to Salyut 6 but with docking facilities to take vehicles weighing 30 000 lb, such as the Star module. Later missions saw the assembly of further solar panels by spacewalking cosmonauts. In March 1985 the Soviets announced that Salyut 7 had ceased manned space operations.

Launched 19 April 1982

Visiting craft

Soyuz T 5	
Progress 13	
Soyuz T 6	
Progress 14	
Soyuz T 7	
Progress 15	
Progress 16	
Star module Cosmos 1443	(unmanned)
Soyuz T 8	Failed to dock
Soyuz T 9	
Progress 17	
Soyuz T-10-1	Launch pad abort
Progress 18	
Soyuz T 10	
Progress 19	
Soyuz T 11	
Progress 20	
Progress 21	
Progress 22	
Soyuz T 12	

The launch of STS 5. (*NASA*)

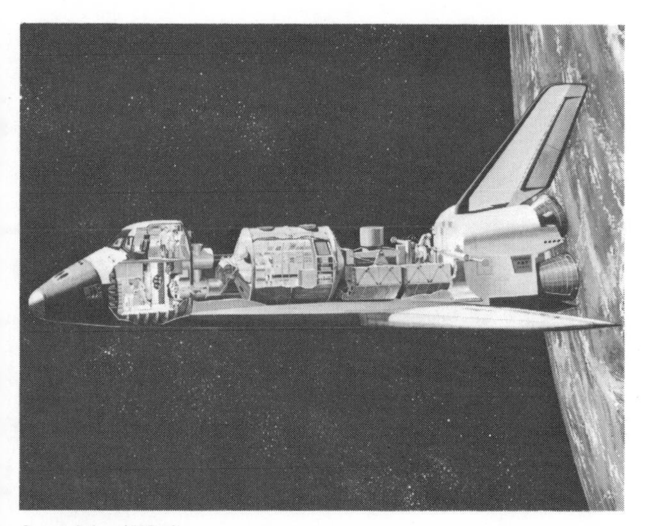

Spacelab. (*ESA*)

Spacelab

Built by the European Space Agency, Spacelab is a modular versatile research centre that flies in the Space Shuttle Orbiter. It comprises any combinations of habitable modules and unpressurized platforms called pallets. The module comprises two 13.1 ft diameter sections – the cone segment and experiment segment. The cone segment can fly on its own in a short module configuration. A Spacelab transfer tunnel, 3.3 ft in diameter, is attached from the airlock to the module. It can be in two lengths, 8.7 ft or 18.8 ft. Spacelab 1 consisted of the long module and a single pallet and weighed 33 584 lb.

Flight log

Spacelab 1–STS 9 28 November 1983 10 days 7 h 47 min 23 s
Spacelab 3–STS 51B 29 April 1985 7 days 0h 8 min 50 s

The landing of STS 9. (*NASA*)

MMU

Manned Manoeuvring Unit (MMU), enabling spacewalking astronauts to fly independently of the spacecraft without tethers. Basically a backpack with arms, the Shuttle MMU weighs 340 lb and is equipped with two nitrogen gas tanks containing 13 lb at 3000 psi. Thrust from 24 pressurized nitrogen thruster jets with a thrust of 1.6 lb are controlled by the right-hand control for pitch, yaw and roll and left-hand, for straight-line motion – up, down, forward, backward via a microprocessor. A total of 729 command combinations are possible. Two MMUs are carried.

Flight log

STS 41B	5 excursions	1 h 22 min
		1 h 9 min
		47 min
		44 min
		1 h 8 min
STS 41C	2 excursions	42 min
		28 min
STS 51A	2 excursions	

4

An A to Z of Space Travellers

Notes (Up to 6 May 1985)
(Also please refer to Manned Space Flight Tables, see p.108)

Gemini 9 and 10 assignments changed because of deaths of original prime crew of Gemini 9.

The original back-up crew of Apollo 1 was assigned to another flight and a new back-up crew assigned.

The original Apollo 2 flight was cancelled and a new Apollo 2 mission created during reschedules. These are called Apollo 2A and 2B.

Apollo 3 became Apollo 9 then became Apollo 8, the original Apollo 8 becoming Apollo 9.

Russia planned to send a one-cosmonaut crew around the Moon in Zond a number of times in 1968–9.

MOL was a USAF project called Manned Orbital Laboratory. The project was cancelled in 1969.

Not all Russian back-up assignments are known, nor are the whereabouts of many retired cosmonauts.

Apollo 18, 19, 20 were all cancelled. Prime and back-up crews had been unofficially assigned to these missions. The original Apollo 17 back-up crew was replaced by new back-ups.

The US ASTP mission was unofficially termed Apollo 18.

Skylab 1 was the space station.

OFT was the original title of the first orbital flight tests of the Space Shuttle, standing for Orbital Flight Tests. Four missions were assigned crews but OFT became STS and crew assignments were altered mainly due to one resignation and the delays to the programme.

ALT stands for Approach and Landing Tests, five manual guide flights by Orbiter Enterprise from the back of a 747 jet to a landing at Edwards AFB in 1977, to initiate the Space Shuttle programme.

The STS numbering system was changed in 1984 to an extremely confusing one. The 'year' ran from October to October, which was the US fiscal year. So 41B was the second mission (B) of year 1984 (4) from Kennedy (1). 62A is the first mission (A) in 1986 (6) from Vandenberg (2).

Also the designations of shuttle flights were altered regularly. Only the latest STS designation has been assigned to each crewman. For example, one astronaut (Bobko) was assigned to STS 14 which became 41E, then 41F and was then cancelled. He was reassigned to 51E which was also cancelled. He then became 51D commander. STS 10 became 41E and then 51C. It would be more unhelpful than educational to include every flight number change. For a complete list of shuttle flight designations, assignments and reassignments please refer to the Manned Space Flight Tables under NASA Flight Crew Selections.

For planned flights of the shuttle, flight dates have not been given. See Space Shuttle Schedule in Manned Space Flight Tables.

Intercosmos research cosmonauts were those from guest countries flying on Soyuz. Intercosmos is a Soviet organization promoting space co-operation within Soviet bloc and other countries.

Although the aborted missions of the original Soyuz 18 and T10 flights are often termed 18A and T-10A and later flights 18B and T-10B, the Soviets in fact term them 18-1, T-10-1 and 18 and T-10.

NAME	Vladimir Viktorovich Aksyonov
SPACE PERSON	79th
SERVICE RANK	Civilian
CURRENT STATUS	Works at Star City, Cosmonaut Training Centre
BIRTH DATE	1 February 1935
BIRTHPLACE	Biglitsy, USSR
EXPERIENCE	Air Force officer Aerospace designer
SPACE CAREER	Cosmonaut, 1973 Soyuz 22 Flight engineer, 15 September 1976 Soyuz T-2 Flight engineer, 5 June 1980 Retired.
MARITAL STATUS	Married, two children
SPACE EXPERIENCE	11 days 20 h 11 min

NAME	Edwin Eugene 'Buzz' Aldrin Jr
SPACE PERSON	30th
SERVICE RANK	Col., US Air Force, retired
CURRENT STATUS	Director Astronomical Programmes, University of North Dakota
BIRTH DATE	20 January 1930
BIRTHPLACE	Monclair, New Jersey, USA
EXPERIENCE	Science degree USAF pilot Korean War veteran 66 combat missions Doctorate in astronautics
SPACE CAREER	NASA Group 3, October 1963 Original Gemini 10 back-up pilot Gemini 9 back-up pilot Gemini 12 pilot, 11 November 1966 Apollo 9 back-up lunar module pilot (rescheduled) Apollo 8 back-up command module pilot Apollo 11 lunar module pilot, 16 July 1969 Resigned, 1971
MARITAL STATUS	Married, 3 children, divorced, remarried
SPACE EXPERIENCE	12 days 1 h 53 min 6 s

NAME	Aleksander Aleksandrov
SPACE PERSON	123rd
SERVICE RANK	Civilian
CURRENT STATUS	Cosmonaut
BIRTH DATE	20 February 1943
BIRTHPLACE	Moscow, USSR
EXPERIENCE	Soviet Army Spacecraft control systems designer
SPACE CAREER	Cosmonaut, 1978 Soyuz T-8 back-up flight engineer Soyuz T-9 flight engineer, 27 June 1983
MARITAL STATUS	Married, 1 child
SPACE EXPERIENCE	149 days 10 h 46 min

NAME	Joseph Percival Allen
SPACE PERSON	112th (joint)
SERVICE RANK	Civilian
CURRENT STATUS	NASA astronaut mission specialist
BIRTH DATE	27 June 1937
BIRTHPLACE	Crawfordsville, Indiana, USA
EXPERIENCE	Maths/physics degree Physics degree Doctorate in physics
SPACE CAREER	NASA Group 6, August 1967 Apollo 15 support crew/mission scientist *Columbia* – STS 5 mission specialist, 11 November 1982 *Discovery* – STS 51A mission specialist, 8 November 1984
MARITAL STATUS	Married, 2 children
SPACE EXPERIENCE	13 days 2 h 0 min 20 s

NAME	William Alison Anders
SPACE PERSON	34th
SERVICE RANK	Major General, US Air Force, Reserve
CURRENT STATUS	Senior Executive, Vice President, Operations, Textron Inc.
BIRTH DATE	17 October 1933
BIRTHPLACE	Hong Kong
EXPERIENCE	Science degree US Air Force pilot Nuclear engineering degree
SPACE CAREER	NASA Group 3, October 1963 Gemini 11 back-up pilot Apollo 3 lunar module pilot (rescheduled) Apollo 9 lunar module pilot (rescheduled) Apollo 8 pilot, 21 December 1968 Apollo 11 back-up command module pilot Resigned, 1969
MARITAL STATUS	Married, 6 children
SPACE EXPERIENCE	6 days 3 h 0 min 42 s

Aleksandrov. *Novosti*

NAME	Neil Alden Armstrong
SPACE PERSON	25th (joint)
SERVICE RANK	Civilian
CURRENT STATUS	Chairman, Computing Technologies for Aviatiar Inc.
BIRTH DATE	5 August 1930
BIRTHPLACE	Wapakoneta, Ohio, USA
EXPERIENCE	US Navy pilot 78 combat missions, Korea Shot down NACA test pilot NASA test pilot X-15 test pilot (7 flights)
SPACE CAREER	NASA Group 2, September 1962 Gemini 5 back-up command pilot Gemini 8 command pilot, 16 March 1966 Gemini 11 back-up command pilot Apollo 9 back-up commander (rescheduled) Apollo 8 back-up commander Apollo 11 commander, 16 July 1969 Resigned, 1971
MARITAL STATUS	Married, 2 children
SPACE EXPERIENCE	8 days 14 h 0 min 1 s

NAME	Yuri Petrovich Artyukhin
SPACE PERSON	71st
SERVICE RANK	Col. Eng, Soviet Air Force
CURRENT STATUS	Unknown (non-active)
BIRTH DATE	22 July 1930
BIRTHPLACE	Pershutino, Moscow, USSR
EXPERIENCE	Air Force Technical College Soviet Air Force
SPACE CAREER	Cosmonaut, January 1963 Soyuz 14 flight engineer, 3 July 1974
MARITAL STATUS	Married, 2 children
SPACE EXPERIENCE	15 days 17 h 30 min

NAME	Oleg Atkov
SPACE PERSON	136th (joint)
SERVICE RANK	Civilian, MD
CURRENT STATUS	Cosmonaut
BIRTH DATE	9 May 1949
BIRTHPLACE	Khvorostyanka, USSR
EXPERIENCE	Medical college graduate Intern Research cardiologist Developed portable ultrasound cardiograph Leninst Kosomol prize
SPACE CAREER	Cosmonaut, 1977 Soyuz T-10 research engineer, 8 February 1984
MARITAL STATUS	Unknown
SPACE EXPERIENCE	236 days 22 h 50 min

NAME	Alan Lavern Bean
SPACE PERSON	45th
SERVICE RANK	Capt., US Navy, retired

Armstrong. *NASA*

CURRENT STATUS	Professional artist
BIRTH DATE	15 March 1932
BIRTHPLACE	Wheeler, Texas, USA
EXPERIENCE	Aeronautical engineering degree US Navy Test pilot
SPACE CAREER	NASA Group 3, October 1963 Gemini 10 back-up command pilot Apollo 8 back-up lunar module pilot (rescheduled) Apollo 9 back-up lunar module pilot Apollo 12 lunar module pilot, 14 November 1969 Skylab 3 commander, 28 July 1973 ASTP back-up commander Resigned, 1981
MARITAL STATUS	Married, 2 children, divorced
SPACE EXPERIENCE	69 days 15 h 45 min 29 s

NAME	Pavel Ivanovich Belyayev, deceased
SPACE PERSON	16th (joint)
SERVICE RANK	Col., Soviet Air Force
CURRENT STATUS	Died natural causes, 10 January 1970
BIRTH DATE	26 June 1925
BIRTHPLACE	Chelishchevo, Vologda, USSR
EXPERIENCE	Mill worker Soviet Air Force pilot Action in Japan, Second World War

SPACE CAREER	Cosmonaut, March 1960 Voskhod 2, commander, 18 March 1965 First Zond pilot, 1967 (mission cancelled)
MARITAL STATUS	Married, 2 children
SPACE EXPERIENCE	1 day 2 h 2 min 17 s

NAME	Georgi Timofeyevich Beregovoi
SPACE PERSON	33rd
SERVICE RANK	Lt Gen., Soviet Air Force
CURRENT STATUS	Commander, Cosmonaut Training Centre, Star City
BIRTH DATE	15 April 1921
BIRTHPLACE	Fyodorovka, Poltava, Ukraine, USSR
EXPERIENCE	Soviet Air Force 185 combat missions, Second World War Hero of Soviet Union Test pilot Merited Test Pilot of USSR award
SPACE CAREER	Cosmonaut, February 1964 Voskhod 3 back-up commander (cancelled) Soyuz 3 pilot, 26 October 1968
MARITAL STATUS	Married, 2 children
SPACE EXPERIENCE	3 days 22 h 51 min

Beregovoi. *Author*

NAME	Anatoli Berezovoi
SPACE PERSON	107th
SERVICE RANK	Col., Soviet Air Force
CURRENT STATUS	Cosmonaut
BIRTH DATE	11 April 1942
BIRTHPLACE	Oktyabrsky, USSR
EXPERIENCE	Lathe operator Military High School for Test Pilots
SPACE CAREER	Cosmonaut, 1970 Soyuz 24 back-up commander Soyuz T-5 commander, 13 May 1982 Soyuz T-11 back-up commander
MARITAL STATUS	Unknown
SPACE EXPERIENCE	211 days 8 h 5 min

NAME	Guion Stewart Bluford
SPACE PERSON	124th (joint)
SERVICE RANK	Col., US Air Force
CURRENT STATUS	NASA astronaut mission specialist
BIRTH DATE	22 November 1942
BIRTHPLACE	Philadelphia, Penn, USA
EXPERIENCE	Aerospace engineering degrees Doctorate aerospace engineering US Air Force Chief Aerodynamics and Airframe Branch Aerodynamics Division, Wright Patterson AFB
SPACE CAREER	NASA Group 8, 1978 *Challenger* – STS 8 mission specialist, 30 August 1983 STS 61A/Spacelab D-1 mission specialist
MARITAL STATUS	Married, 2 children
SPACE EXPERIENCE	6 days 1 h 8 min 40 s

Bluford. *NASA*

NAME	Karol Joseph Bobko
SPACE PERSON	115th (joint)
SERVICE RANK	Col., US Air Force
CURRENT STATUS	NASA astronaut commander
BIRTH DATE	23 December 1937
BIRTHPLACE	New York City, USA
EXPERIENCE	Science degree US Air Force pilot
SPACE CAREER	USAF MOL Group 2, June 1966 NASA Group 7, August 1969 ASTP support crew *Challenger* – STS 6 pilot, 4 April 1983 *Discovery* STS 51D commander, 12 April, 1985 STS 51J commander
MARITAL STATUS	Married, 2 children
SPACE EXPERIENCE	12 days 0 h 19 min 42 s

NAME	Frank Borman
SPACE PERSON	22nd (Joint)
SERVICE RANK	Col., US Air Force, retired
CURRENT STATUS	Chairman, Eastern Air Lines
BIRTH DATE	14 March 1928
BIRTHPLACE	Gary, Indiana, USA
EXPERIENCE	Science degree US Air Force pilot Taught thermodynamics and fluid mechanics at US Military Academy Aeronautical engineering Aerospace research pilot Aerospace research pilot instructor, Edwards AFB
SPACE CAREER	NASA Group 2, September 1962 Gemini 4 back-up command pilot Gemini 7 commander, 4 December 1965 Apollo 2A commander (cancelled) Apollo 3 commander (redesignated as) Apollo 9 commander (rescheduled) Apollo 8 commander, 21 December 1968 Resigned, 1970
MARITAL STATUS	Married, 2 children
SPACE EXPERIENCE	19 days 21 h 35 min 33 s

NAME	Vance DeVoe Brand
SPACE PERSON	76th (joint)
SERVICE RANK	Civilian
CURRENT STATUS	NASA astronaut commander
BIRTH DATE	9 May 1931
BIRTHPLACE	Longmont, Colorado, USA
EXPERIENCE	Business administration degree Aeronautical engineering degree US Marine Corp Test pilot, Lockheed Aircraft Corp
SPACE CAREER	NASA Group 5, April 1966 Apollo 8 support crew Apollo 13 support crew Apollo 15 back-up command module pilot Apollo 18 command module pilot (cancelled) Skylab 3 back-up commander Skylab 4 back-up commander Skylab Rescue Standby commander ASTP command module pilot, 15 July 1975 OFT-4 commander (reassigned) *Columbia* – STS 5 commander, 11 November 1982 *Challenger* – STS 41B commander, 3 February 1984 STS 61D Spacelab 4 (SLS-1) commander
MARITAL STATUS	Married, 4 children, divorced, remarried, 1 child
SPACE EXPERIENCE	22 days 2 h 58 min 44 s

Borman. *Eastern Airlines*

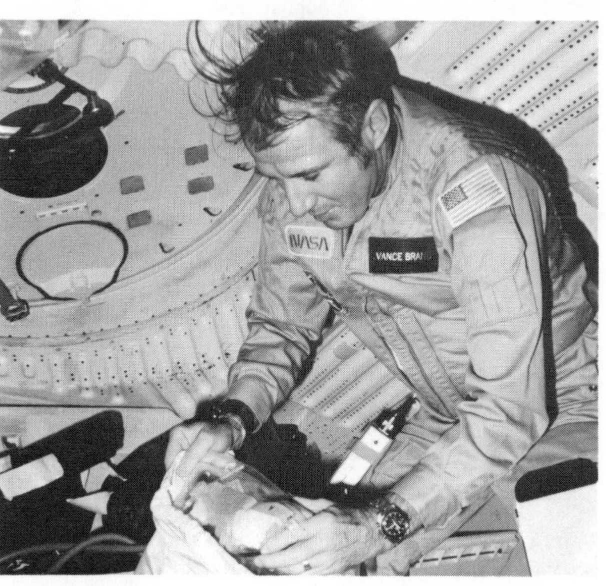

Brand. *NASA*

NAME	Daniel Charles Brandenstein
SPACE PERSON	124th (joint)
SERVICE RANK	Capt., US Navy
CURRENT STATUS	NASA astronaut commander
BIRTH DATE	17 January 1943
BIRTHPLACE	Waterton, Wisconsin, USA
EXPERIENCE	Physics and maths degree US Navy Vietnam veteran 192 combat missions Officer, Attack Squadron 145
SPACE CAREER	NASA Group 8, 1978 *Challenger* STS 8 pilot, 30 August 1983 STS 51G commander
MARITAL STATUS	Married, 1 child
SPACE EXPERIENCE	6 days 1 h 8 min 40 s

NAME	James Frederick Buchli
SPACE PERSON	156th (joint)
SERVICE RANK	Lt Col., US Marine Corps
CURRENT STATUS	NASA astronaut mission specialist
BIRTH DATE	20 June 1945
BIRTHPLACE	New Rockford, South Dakota, USA
EXPERIENCE	Science degrees USMC pilot
SPACE CAREER	NASA Group 8, August 1978 *Discovery* – STS 51C mission specialist, 24 January 1985 STS 61A Spacelab D-1 mission specialist
MARITAL STATUS	Married, 2 children
SPACE EXPERIENCE	3 days, 1 h 33 min 13 s

NAME	Valeri Fyodorovich Bykovsky
SPACE PERSON	11th
SERVICE RANK	Col., Soviet Air Force
CURRENT STATUS	Space official, Star City
BIRTH DATE	2 August 1934
BIRTHPLACE	Pavlov-Posad, Moscow, USSR
EXPERIENCE	Learned to fly at 17 Soviet Air Force
SPACE CAREER	Cosmonaut, March 1960 Vostok 3 back-up pilot Vostok 5 pilot, 14 June 1963 Soyuz 2 commander, mission cancelled Zond pilot, mission cancelled Soyuz 22 commander, 15 September 1976 Soyuz 31 commander, 26 August 1978 Soyuz 37 back-up commander Retired
MARITAL STATUS	Married, 1 child
SPACE EXPERIENCE	20 days 17 h 47 min

NAME	Malcolm Scott Carpenter
SPACE PERSON	6th
SERVICE RANK	Commander, US Navy, retired
CURRENT STATUS	President, Sea Services Inc.
BIRTH DATE	1 May 1925
BIRTHPLACE	Boulder, Colorado, USA
EXPERIENCE	Aeronautical engineering US Navy, pilots wings Anti-sub patrol duties, Korea Intelligence officer, USS *Hornet*
SPACE CAREER	NASA Group 1, April 1959 MA-6 back-up pilot MA-7 pilot, 24 May 1962 Resigned, 1967
MARITAL STATUS	Married, 4 children, divorced, remarried, 2 children
SPACE EXPERIENCE	4 h 56 min 5 s

NAME	Gerald Paul Carr
SPACE PERSON	66th (joint)
SERVICE RANK	Col., US Marine Corps, retired
CURRENT STATUS	Senior Consultant, Applied Research Inc.
BIRTH DATE	22 August 1932
BIRTHPLACE	Denver, Colorado, USA
EXPERIENCE	Mechanical engineering degree Aeronautical engineering degree US Navy US Marines Test pilot
SPACE CAREER	NASA Group 5, 1966 Apollo 8 support crew Apollo 12 support crew Original Apollo 16 back-up lunar module pilot Apollo 19 lunar module pilot (cancelled) Skylab 4 commander, 16 November 1973
MARITAL STATUS	Married, 6 children
SPACE EXPERIENCE	84 days 1 h 15 min 31 s

NAME	Eugene Andrew Cernan
SPACE PERSON	27th
SERVICE RANK	Capt., US Navy, retired
CURRENT STATUS	President, The Cernan Energy Group
BIRTH DATE	14 March 1934
BIRTHPLACE	Chicago, Illinois, USA
EXPERIENCE	Electrical engineering degree US Navy Aeronautical engineering degree
SPACE CAREER	NASA Group 3, October 1963 Gemini 9 back-up pilot (prime crew killed) Gemini 9 pilot, 3 June 1966 Gemini 12 back-up pilot Apollo 2B back-up lunar module pilot (cancelled) Apollo 7 back-up pilot Apollo 10 lunar module pilot, 18 May 1969 Apollo 14 back-up commander Apollo 17 commander, 8 December 1972 Resigned, 1976
MARITAL STATUS	Married, 1 child
SPACE EXPERIENCE	23 days 14 h 16 min 12 s

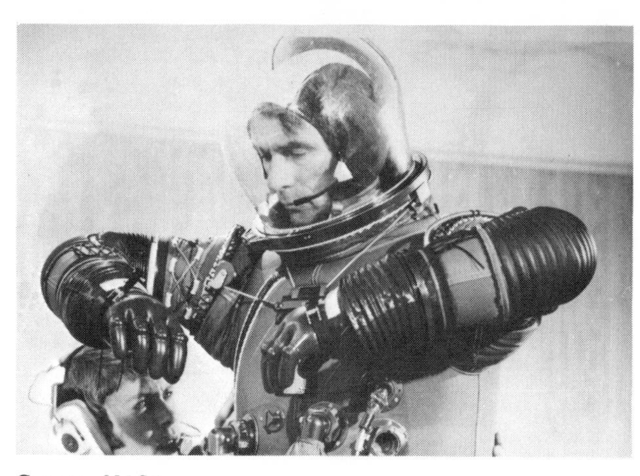

Cernan. *NASA*

NAME	Jean-Loup Chrétien
SPACE PERSON	108th
SERVICE RANK	Lt Col., French Air Force
CURRENT STATUS	CNES astronaut
BIRTH DATE	20 August 1938
BIRTHPLACE	La Rochelle, France
EXPERIENCE	Chief test pilot, Mirage F-1 fighter
SPACE CAREER	Intercosmos candidate, 1980 Soyuz T-6 cosmonaut researcher, 24 June 1982 STS 51G back-up
MARITAL STATUS	Married, 4 children
SPACE EXPERIENCE	7 days 22 h 42 min

Chrétien. *CNES*

NAME	Michael Loyd Coats
SPACE PERSON	144th (joint)
SERVICE RANK	Commander, US Navy
CURRENT STATUS	NASA astronaut commander
BIRTH DATE	16 January 1946
BIRTHPLACE	Sacramento, California, USA
EXPERIENCE	Science degree US Navy pilot Vietnam veteran 315 combat missions
SPACE CAREER	NASA Group 8, August 1978 *Discovery* – STS 41D pilot, 30 August 1984 STS 61H commander
MARITAL STATUS	Married, 1 child
SPACE EXPERIENCE	6 days 0 h 56 min 4 s

NAME	Michael Collins
SPACE PERSON	28th
SERVICE RANK	Maj. Gen., US Air Force (reserve)
CURRENT STATUS	Vice President, Vought Corporation
BIRTH DATE	31 October 1930
BIRTHPLACE	Rome, Italy
EXPERIENCE	Science degree Test pilot, Edwards AFB
SPACE CAREER	NASA Group 3, October 1963 Gemini 7 back-up pilot Gemini 10 pilot, 18 July 1966 Apollo 2A back-up pilot (cancelled) Apollo 3 command module pilot (rescheduled) Apollo 9 command module pilot (rescheduled) Apollo 8 command module pilot Dropped from flight status due to spinal injury Apollo 11 command module pilot, 16 July 1969 Resigned, 1970
MARITAL STATUS	Married, 3 children
SPACE EXPERIENCE	11 days 2 h 5 min 14 s

NAME	Charles 'Pete' Conrad
SPACE PERSON	21st
SERVICE RANK	Capt., US Navy, retired
CURRENT STATUS	Corporate Vice President, McDonnell Douglas Corporation
BIRTH DATE	2 June 1930
BIRTHPLACE	Philadelphia, Pennsylvania, USA
EXPERIENCE	Aeronautical engineering degree US Navy test-pilot school Flight instructor, project test pilot
SPACE CAREER	Group 2 NASA astronaut, 1962 Gemini 5, pilot 21 August 1965 Gemini 8, back-up commander Gemini 11, command pilot 11 September 1966 Apollo 3 back-up commander (rescheduled) Apollo 8 back-up commander (rescheduled) Apollo 9 back-up commander Apollo 12 commander, 12 November 1969 Apollo 17 original back-up commander Apollo 20 commander (cancelled) Skylab 2 commander, 25 May 1973 Resigned, 1974
MARITAL STATUS	Married, 4 children
SPACE EXPERIENCE	49 days 3 h 38 min 36 s

Conrad. *McDonnell Douglas*

NAME	Leroy Gordon Cooper
SPACE PERSON	10th
SERVICE RANK	Col., US Air Force, retired
CURRENT STATUS	President, Vistec
BIRTH DATE	6 March 1927
BIRTHPLACE	Shawnee, Oklahoma, USA
EXPERIENCE	Aeronautical engineering degree US Marines US Navy US Army US Air Force Test pilot instructor, Edwards AFB
SPACE CAREER	NASA Group 1, 1959 MA8 back-up pilot MA9 pilot, 15 May 1963 Gemini 5 command pilot, 21 August 1965 Gemini 12 back-up command pilot Apollo 10 back-up commander Resigned, 1970
MARITAL STATUS	Married, 2 children, divorced, remarried
SPACE EXPERIENCE	9 days 9 h 15 min 3 s

NAME	Robert Laurel 'Crip' Crippen
SPACE PERSON	102nd
SERVICE RANK	Capt., US Navy
CURRENT STATUS	NASA astronaut commander
BIRTH DATE	11 September 1937
BIRTHPLACE	Beaumont, Texas, USA
EXPERIENCE	Aerospace engineering degree US Navy Aerospace research pilot ARP Instructor
SPACE CAREER	USAF MOL Group 2, June 1966 NASA Group 7, August 1969 Skylab 2 support crew Skylab 3 support crew Skylab 4 support crew ASTP support crew leader OFT 1 pilot (redesignated) *Columbia* – STS1 pilot, 12 April 1981 *Challenger* – STS7 commander, 18 June 1983 *Challenger* – STS 41C commander, 6 April 1984 *Challenger* – STS 41G commander, 5 October 1984 STS 62A commander
MARITAL STATUS	Married, 3 children
SPACE EXPERIENCE	23 days 13 h 48 min 40 s

Crippen. *NASA*

NAME	Walter Cunningham
SPACE PERSON	31st (joint)
SERVICE RANK	Civilian
CURRENT STATUS	Vice President, the Capital Group
BIRTH DATE	16 March 1932
BIRTHPLACE	Creston, Iowa, USA
EXPERIENCE	US Navy pilot Two science degrees Doctorate in physics Research scientist, Rand Corporation
SPACE CAREER	NASA Group 3, October 1963 Apollo 2A pilot (cancelled) Apollo 1 back-up pilot Apollo 7 pilot, 11 October 1968 Original Skylab commander (Replaced by redundant Apollo crews) Resigned, 1971
MARITAL STATUS	Married, 2 children
SPACE EXPERIENCE	10 days 20 h 9 min 3 s

NAME	Lev Stepanovich Demin
SPACE PERSON	72nd (joint)
SERVICE RANK	Col., Soviet Air Force
CURRENT STATUS	Chief Soviet Stamp Federation
BIRTH DATE	11 January 1926
BIRTHPLACE	Moscow, USSR
EXPERIENCE	Graduate Zhukovsky Engineering Academy
SPACE CAREER	Cosmonaut, January 1963 Soyuz 15 flight engineer, 25 August 1974 Retired
MARITAL STATUS	Married, 2 children
SPACE EXPERIENCE	2 days 0 h 12 min

NAME	Georgi Timofeyevich Dobrovolsky, deceased
SPACE PERSON	52nd (joint)
SERVICE RANK	Lt Col., Soviet Air Force
CURRENT STATUS	Died during Soyuz 11 descent, 29 June 1971
BIRTH DATE	1 June 1928
BIRTHPLACE	Odessa, USSR
EXPERIENCE	Air Force School Air Force Academy
SPACE CAREER	Cosmonaut, January 1963 Soyuz 10 back-up commander 6 June 1971 Soyuz 11 commander
MARITAL STATUS	Married, 2 children
SPACE EXPERIENCE	23 days 18 h 22 min

Dobrovolsky. *Novosti*

NAME	Charles Moss Duke Jr
SPACE PERSON	56th (joint)
SERVICE RANK	Brig. Gen., US Air Force, reserve
CURRENT STATUS	President Southwest Wilderness Art Inc. Duke Investments
BIRTH DATE	3 October 1935
BIRTHPLACE	Charlotte, North Carolina, USA
EXPERIENCE	Naval science degree Astronautics and aeronautics degree from US Naval Academy USAF Aerospace Research Pilot (ARP) ARP Instructor
SPACE CAREER	NASA Group 5, April 1966 Apollo 10 support crew Apollo 11 support crew (landing capsule communicator) Apollo 13 back-up lunar module pilot Apollo 16 lunar module pilot, 16 April 1972 Apollo 17 back-up lunar module pilot Resigned, 1976
MARITAL STATUS	Married, 2 children
SPACE EXPERIENCE	11 days 1 h 5 min 5 s

NAME	Vladimir Alexandrovich Dzhanibekov
SPACE PERSON	86th
SERVICE RANK	Col., Soviet Air Force
CURRENT STATUS	Cosmonaut
BIRTH DATE	13 May 1942
BIRTHPLACE	Iskander, USSR
EXPERIENCE	Physics diploma Flying Club pilot Military flying school Radio and electronics expert
SPACE CAREER	Cosmonaut 1970 Soyuz 19 back-up commander Soyuz 27 commander, 10 January 1978 Soyuz 36 back-up commander Soyuz 39 commander, 22 March 1981 Soyuz T-6 commander, 24 June 1982 Soyuz T-12 commander, 17 July 1984
MARITAL STATUS	Married, 2 children
SPACE EXPERIENCE	33 days 13 h 38 min

Dzhanibekov. *Novosti*

NAME	Donn Fulton Eisele
SPACE PERSON	31st (joint)
SERVICE RANK	Col., US Air Force, retired
CURRENT STATUS	Oppenheimer and Co Inc.
BIRTH DATE	23 June 1930
BIRTHPLACE	Columbus, Ohio, USA
EXPERIENCE	Science degree US Naval Academy US Air Force pilot Astronautics degree Aerospace research pilot
SPACE CAREER	NASA Group 3, October 1963 Original Apollo 1 pilot, injured shoulder, dropped Apollo 2A senior pilot (cancelled) Apollo 1 back-up senior pilot Apollo 7 senior pilot, 11 October 1968 Apollo 10, back-up command module pilot Resigned, 1972
MARITAL STATUS	Married, 4 children, divorced, remarried, 1 child
SPACE EXPERIENCE	10 days 20 h 9 min 3 s

NAME	Joseph Henry Engle
SPACE PERSON	104th (joint)
SERVICE RANK	Col., US Air Force
CURRENT STATUS	NASA astronaut commander
BIRTH DATE	26 August 1932
BIRTHPLACE	Abilene, Kansas, USA
EXPERIENCE	Aeronautical engineering degree, US Air Force, 1955 US experimental test pilot Aerospace research pilot X-15 pilot 16 flights, including 3 'astroflights' US Air Force Astronaut Wings
SPACE CAREER	NASA Group 5, April 1966 Apollo 10 support crew Apollo 14 back-up lunar module pilot Apollo 17 lunar module pilot, dropped *Enterprise*–Shuttle ALT commander, 1977 OFT commander (redesignated) *Columbia* – STS1 back-up commander STS2 commander, 12 November 1981 STS 51I commander
MARITAL STATUS	Married, 2 children
SPACE EXPERIENCE	2 days 6 h 13 min 11 s

NAME	Ronald Ellwin Evans
SPACE PERSON	58th (joint)
SERVICE RANK	Capt., US Navy, retired
CURRENT STATUS	Manager, Space Products, Sperry Flight Systems
BIRTH DATE	10 November 1933
BIRTHPLACE	St Francis, Kansas, USA
EXPERIENCE	Electrical engineering degree Aeronautical engineering degree US Navy Vietnam veteran Flew F-8 fighters from USS *Ticonderoga*
SPACE CAREER	NASA Group 5, April 1966 Apollo 1 support crew Apollo 7 support crew Apollo 11 support crew Apollo 14 back-up command module pilot Apollo 17 command module pilot, 7 December 1972 ASTP back-up command module pilot Resigned, 1977
MARITAL STATUS	Married, 2 children
SPACE EXPERIENCE	12 days 13 h 51 min 59 s

NAME	John McCreary Fabian
SPACE PERSON	119th (joint)
SERVICE RANK	Col., US Air Force
CURRENT STATUS	NASA astronaut mission specialist pilot
BIRTH DATE	28 January 1939
BIRTHPLACE	Goosecreek, Texas, USA
EXPERIENCE	Mechanical engineering degree Aerospace engineering degree Doctorate in aeronautics and astronautics US Air Force pilot Vietnam veteran 90 combat missions Assistant Professor of Aeronautics USAF Academy
SPACE CAREER	NASA Group 8, August 1978 *Challenger* – STS 7 mission specialist, 18 June 1983 STS 51G mission specialist STS 61D/Spacelab 4 (SLS-1) mission specialist
MARITAL STATUS	Married, 2 children
SPACE EXPERIENCE	6 days 2 h 24 min 10 s

NAME	Bertalan Farkas
SPACE PERSON	94th
SERVICE RANK	Lt Col., Hungarian Air Force
CURRENT STATUS	Unknown (non-active)
BIRTH DATE	2 August 1949
BIRTHPLACE	Gyulahaza, Hungary
EXPERIENCE	Glider pilot Aeronautical engineering degree Army Air Force
SPACE CAREER	Intercosmos candidate, 1978 Soyuz 36 cosmonaut researcher, 26 May 1980 Retired
MARITAL STATUS	Married, 2 children
SPACE EXPERIENCE	7 days 20 h 46 min

NAME	Konstantin Petrovich Feoktistov
SPACE PERSON	13th (joint)
SERVICE RANK	Civilian
CURRENT STATUS	Senior scientist, manned spaceflight
BIRTH DATE	26 February, 1926
BIRTHPLACE	Voronezh, USSR
EXPERIENCE	Studied engineering in Moscow Spacecraft designer
SPACE CAREER	Cosmonaut 1964 Voskhod 1 science pilot, 12 October 1964 Original Soyuz T-3 research engineer, dropped due to physical problems
MARITAL STATUS	Married, 1 child
SPACE EXPERIENCE	1 day 0 h 17 min 3 s

Feoktistov. *Novosti*

NAME	Anatoli Vasilyevich Filipchenko
SPACE PERSON	42nd (joint)
SERVICE RANK	Maj. Gen., Soviet Air Force
CURRENT STATUS	Unknown (non-active)
BIRTH DATE	26 February 1928
BIRTHPLACE	Davydovka, USSR
EXPERIENCE	Lathe operator Military school Air Force
SPACE CAREER	Cosmonaut, January 1963 Soyuz 4 back-up commander Soyuz 7 commander, 12 October 1969 Soyuz 9 back-up commander Soyuz 16 commander, 2 December 1974 Retired
MARITAL STATUS	Married, 2 children
SPACE EXPERIENCE	10 days 23 h 5 min

Filipchenko. *Novosti*

NAME	Anna Lee Fisher
SPACE PERSON	154th (joint)
SERVICE RANK	Civilian, MD
CURRENT STATUS	NASA astronaut mission specialist
BIRTH DATE	24 August 1949
BIRTHPLACE	Albany, New York, USA
EXPERIENCE	Chemistry degree Doctorate in medicine
SPACE CAREER	NASA Group 8, August 1978 *Discovery* – STS 51A mission specialist, 8 November 1984 STS 61H mission specialist
MARITAL STATUS	Married (to NASA astronaut Bill Fisher), 1 child
SPACE EXPERIENCE	7 days 23 h 45 min 54 s

NAME	Charles Gordon Fullerton
SPACE PERSON	106th
SERVICE RANK	Col., US Air Force
CURRENT STATUS	NASA astronaut commander
BIRTH DATE	1 October 1936
BIRTHPLACE	Rochester, New York, USA
EXPERIENCE	Science degree Mechanical engineering degree Hughes Aircraft Corp US Air Force Pilots Wings B-47 pilot SAC Aerospace research pilot
SPACE CAREER	USAF MOL Group 2, June 1966 NASA Group 7, August 1969 Apollo 14 support crew Apollo 17 support crew *Enterprise*-Shuttle ALT pilot, 1977 OFT 4 pilot (redesignated) *Columbia* – STS 3 pilot, 22 March 1982 STS 51F Spacelab 2 commander

MARITAL STATUS	Married, 2 children
SPACE EXPERIENCE	8 days 0 h 4 min 46 s

Fullerton. *NASA*

NAME	Yuri Alexeyevich Gagarin, deceased
SPACE PERSON	1st
SERVICE RANK	Col., Soviet Air Force Died in air crash, 27 March 1968
BIRTH DATE	9 March 1934
BIRTHPLACE	Klushino, near Smolensk, USSR
EXPERIENCE	Pilot Parachutist Soviet Air Force jet fighter pilot
SPACE CAREER	Cosmonaut group, March 1960 Vostok 1 pilot, 12 April 1961 Soyuz 1 back-up commander Selected for original Soyuz 3
MARITAL STATUS	Married, 2 children
SPACE EXPERIENCE	1 h 48 min

NAME	Dale Allen Gardner
SPACE PERSON	124th (joint)
SERVICE RANK	Commander US Navy
CURRENT STATUS	NASA astronaut mission specialist
BIRTH DATE	8 November 1948
BIRTHPLACE	Fairmont, Minnesota, USA
EXPERIENCE	Engineering physics degree US Navy Flight officer, Air Test and Evaluation Squadron
SPACE CAREER	NASA Group 8, August 1978 *Challenger* – STS 8 mission specialist, 30 August 1983 *Discovery* – STS 51A mission specialist, 8 November 1984 STS 62A mission specialist
MARITAL STATUS	Married, 1 child
SPACE EXPERIENCE	14 days 0 h 54 min 34 s

NAME:	Edwin 'Jake' Garn
SPACE PERSON:	160th (joint)
SERVICE RANK:	Civilian
CURRENT STATUS:	US Senator for Utah
BIRTHDATE:	12 October 1932
BIRTHPLACE:	Richfield, Utah, USA
EXPERIENCE:	Business and finance degree US Navy pilot Insurance executive Mayor Salt Lake City US Senator
SPACE CAREER:	*Discovery* – STS 51D congressional observer, 12 April 1985
MARITAL STATUS:	Married, 4 children, widowed, married, 3 children
SPACE EXPERIENCE:	6 days 23 h 56 min

NAME	Marc Garneau
SPACE PERSON	149th (joint)
SERVICE RANK	Commander, Canadian Navy
CURRENT STATUS	Canadian astronaut
BIRTH DATE	23 February 1949
BIRTHPLACE	Quebec City, Quebec, Canada
EXPERIENCE	Engineering physics degree Doctorate in electrical engineering Combat Systems Engineer, Canadian Navy Project engineer, Naval Weapons Systems Chief Designer Communications and Electronic Warfare Equipment, Canadian Navy
SPACE CAREER	Canadian astronaut, December 1983 *Challenger* – STS 41G payload specialist, 5 October 1984
MARITAL STATUS	Married, 2 children
SPACE EXPERIENCE	8 days 5 h 23 min 33 s

Gagarin. *Novosti*

NAME: Owen Kay Garriott
SPACE PERSON: 62nd (joint)
SERVICE RANK: Civilian
CURRENT STATUS: NASA astronaut mission specialist
BIRTH DATE: 22 November 1930
BIRTHPLACE: Enid, Oklahoma, USA
EXPERIENCE: Two electrical engineering degrees
Doctorate in electrical engineering
Lecturer, Stanford University
SPACE CAREER: NASA Group 4, June 1965
Skylab 3 science pilot, 28 July 1973
Columbia – STS9/Spacelab 1 mission specialist, 28 November 1983
MARITAL STATUS: Married, 4 children
SPACE EXPERIENCE: 69 days 18 h 56 min 27 s

NAME: Edward George Gibson
SPACE PERSON: 66th (joint)
SERVICE RANK: Civilian
CURRENT STATUS: Advanced systems manager, TRW Inc.
BIRTH DATE: 8 November 1936
BIRTHPLACE: Buffalo, New York, USA
EXPERIENCE: Doctorate in engineering physics
SPACE CAREER: NASA Group 4, June 1964
Apollo 12 support crew
Skylab 4 science pilot, 16 November 1973
Resigned, 1974
Rejoined NASA, 1977
Resigned, 1980
MARITAL STATUS: Married, 4 children
SPACE EXPERIENCE: 84 days 1 h 15 min 31 s

NAME: Robert Lee 'Hoot' Gibson
SPACE PERSON: 132nd (joint)
SERVICE RANK: Commander, US Navy
CURRENT STATUS: NASA astronaut commander
BIRTH DATE: 30 October 1946
BIRTHPLACE: Cooperstown, New York, USA
EXPERIENCE: Aeronautical engineering degree
US Navy
Vietnam veteran
56 combat missions
SPACE CAREER: NASA Group 8, August 1978
Challenger – STS 41B pilot, 3 February 1984
STS 61C commander
MARITAL STATUS: Married (to NASA astronaut Rhea Seddon), 1 child
SPACE EXPERIENCE: 7 days 23 h 15 min 54 s

NAME: Yuri Nikolayevich Glazkov
SPACE PERSON: 82nd
SERVICE RANK: Col. Eng., Soviet Air Force
CURRENT STATUS: Senior official, crew training
BIRTH DATE: 2 October 1939
BIRTHPLACE: Moscow, USSR
EXPERIENCE: Air Force Higher Engineering College
SPACE CAREER: Cosmonaut, 1965
Soyuz 23 back-up flight engineer
Soyuz 24 flight engineer, 7 February 1977
Retired
MARITAL STATUS: Married
SPACE EXPERIENCE: 17 days 17 h 26 min

NAME: John Herschel Glenn Jr
SPACE PERSON: 5th
SERVICE RANK: Col., US Marine Corps, retired
CURRENT STATUS: US Democrat Senator, Ohio
BIRTH DATE: 18 July 1921
BIRTHPLACE: Cambridge, Ohio, USA
EXPERIENCE: US Navy Cadet
US Marine Corps pilot
59 combat missions in Pacific, Second World War
63 combat missions Korea
Downed 3 MIG-15s, air medal 18 clusters
Pilot Trans-US Project Bullet
SPACE CAREER: NASA Group 1, April 1959
MR-3 back-up pilot
MR-4 back-up pilot
MR-5 pilot (cancelled)
MA-6 pilot, 20 February 1962
Resigned, 1964
MARITAL STATUS: Married, 2 children
SPACE EXPERIENCE: 4 h 55 min 23 s

NAME: Viktor Vasilyevich Gorbatko
SPACE PERSON: 42nd (joint)
SERVICE RANK: Maj. Gen., Soviet Air Force
CURRENT STATUS: Official, Mission Training, Star City
BIRTH DATE: 3 December 1934
BIRTHPLACE: Kuban, USSR
EXPERIENCE: Soviet Air Force
SPACE CAREER: Cosmonaut, March 1960
Soyuz 2 back-up research engineer (cancelled)
Soyuz 5 back-up research engineer
Soyuz 7 research engineer, 12 October 1969
Soyuz 23 back-up commander
Soyuz 24 commander, 7 February 1977
Soyuz 31 back-up commander
Soyuz 37 commander, 23 July 1980
Retired
MARITAL STATUS: Married, 2 children
SPACE EXPERIENCE: 30 days 12 h 49 min

NAME: Richard Francis Gordon
SPACE PERSON: 29th
SERVICE RANK: Capt., US Navy, retired
CURRENT STATUS: President Astro Systems Engineering Co
BIRTH DATE: 5 October 1929
BIRTHPLACE: Seattle, Washington, USA

EXPERIENCE	Science degree US Navy pilot Project Bullet Trans US Record Winner Bendix trophy Test pilot
SPACE CAREER	NASA Group 3, October 1963 Gemini 8 back-up pilot Gemini 11 pilot, 12 September 1966 Apollo 3 back-up command module pilot (rescheduled) Apollo 8 back-up command module pilot (rescheduled) Apollo 9 back-up command module pilot Apollo 12 command module pilot, 14 November 1969 Apollo 15 back-up commander Apollo 18 commander (cancelled)
MARITAL STATUS	Married, 6 children
SPACE EXPERIENCE	13 days 3 h 53 min 33 s

NAME	Georgi Mikhailovich Grechko
SPACE PERSON	74th (joint)
SERVICE RANK	Civilian
CURRENT STATUS	Unknown (non-active)
BIRTH DATE	25 May 1932
BIRTHPLACE	Leningrad, USSR
EXPERIENCE	Leningrad Institute of Mechanics Lunar spacecraft designer
SPACE CAREER	Cosmonaut, January 1967 Soyuz 7 back-up research engineer Soyuz 9 back-up flight engineer Soyuz 12 back-up flight engineer Soyuz 17 flight engineer, 11 January 1975 Soyuz 26 flight engineer, 10 December 1977 Soyuz T-11 back-up flight engineer Retired
MARITAL STATUS	Married, 2 children
SPACE EXPERIENCE	125 days 23 h 20 min

Grechko. *Novosti*

NAME	Frederick Drew Gregory
SPACE PERSON	165th joint
SERVICE RANK	Col., US Air Force
CURRENT STATUS	NASA astronaut shuttle pilot
BIRTHDATE	7 January 1941
BIRTHPLACE	Washington DC, USA
EXPERIENCE	Science degree US Air Force Helicopter pilot Fighter pilot Test pilot Information systems degree
SPACE CAREER	NASA Group 8, August 1978 *Challenger* – STS 51B Spacelab 3 pilot, 29 April 1985
MARITAL STATUS	Married, two children
SPACE EXPERIECNCE	7 days 0 h 8 min 50 s

NAME	Stanley David Griggs
SPACE PERSON	160th (joint)
SERVICE RANK	Capt., US Navy (Res)
CURRENT STATUS	NASA astronaut pilot/mission specialist
BIRTHDATE	7 September 1939
BIRTHPLACE	Portland, Oregon, USA
EXPERIENCE	Science degree US Navy pilot NASA pilot
SPACE CAREER	NASA Group 8, August 1978 *Discovery* – STS 51D mission specialist, 12 April 1985 *Discovery* – STS 61D pilot
MARITAL STATUS	Married, 2 children
SPACE EXPERIENCE	6 days 23 h 56 mins

NAME	Virgil Ivan 'Gus' Grissom, deceased
SPACE PERSON	3rd
SERVICE RANK	Lt Col., US Air Force Killed in Apollo 1 spacecraft fire, 27 January 1967
BIRTH DATE	23 April 1926
BIRTHPLACE	Mitchell, Indiana, USA
EXPERIENCE	Mechanical engineering US Air Force pilot 100 combat missions, F86 Sabres, Korea Aeronautical degree Test pilot
SPACE CAREER	NASA Group 1, 1959 MR-4 pilot, 21 July 1961 Gemini 3 command pilot, 23 March 1965 Gemini 6, back-up command pilot Apollo 1 commander
MARITAL STATUS	Married, 2 children
SPACE EXPERIENCE	5 h 8 min 28 s

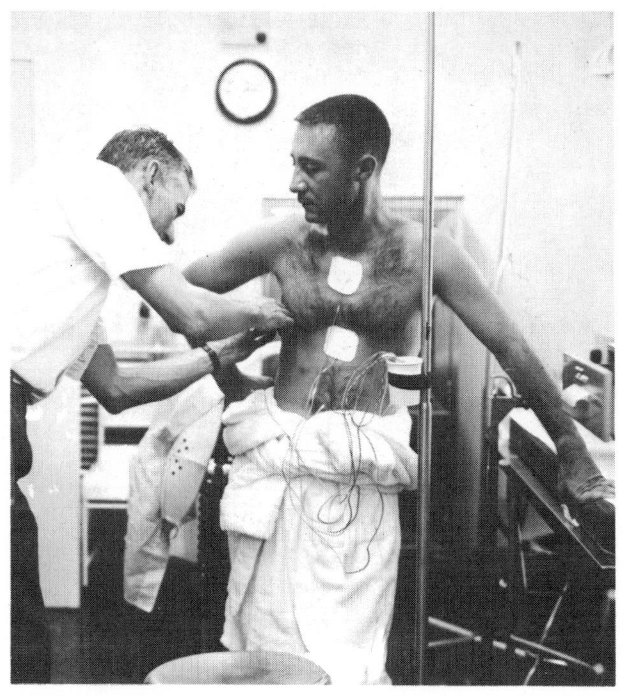

Grissom. *NASA*

NAME	Alexei Alexandrovich Gubarev
SPACE PERSON	74th (joint)
SERVICE RANK	Maj. Gen., Soviet Air Force
CURRENT STATUS	Cosmonaut trainer, Star City
BIRTH DATE	29 April 1932
BIRTHPLACE	Gvardeitsy, Borsky, USSR
EXPERIENCE	Soviet Army Naval Air School Air Force Academy
SPACE CAREER	Cosmonaut, January 1963 Soyuz 12 back-up commander Soyuz 17 commander, 11 January 1975 Soyuz 28 commander, 2 March 1978 Retired
MARITAL STATUS	Married, 1 child
SPACE EXPERIENCE	37 days 7 h 36 min

NAME	Jugderdemidyin Gurragcha
SPACE PERSON	101st
SERVICE RANK	Maj. Gen., Mongolian People's Air Force
CURRENT STATUS	Unknown (non-active)
BIRTH DATE	5 December 1947
BIRTHPLACE	Rashant, Gurvant Settlement, Bulgan Province, Mongolia
EXPERIENCE	Army Aviation School USSR Air Force Academy graduate
SPACE CAREER	Intercosmos candidate, 1978 Soyuz 39 cosmonaut researcher, 22 March 1981 Retired
MARITAL STATUS	Married
SPACE EXPERIENCE	7 days 20 h 43 min

NAME	Fred Wallace Haise
SPACE PERSON	46th (joint)
SERVICE RANK	Civilian
CURRENT STATUS	President, Technical Services Division, Grumman Aerospace
BIRTH DATE	14 November 1933
BIRTHPLACE	Biloxi, Mississippi, USA
EXPERIENCE	Science degree US Navy pilot US Air Force pilot US Marine Corps pilot NASA test pilot
SPACE CAREER	NASA Group 5, April 1966 Apollo 1 support crew Apollo 8 back-up pilot Apollo 11 back-up lunar module pilot Apollo 13 lunar module pilot, 11 April 1970 Apollo 16 back-up commander Apollo 19 commander (cancelled) *Enterprise* – ALT commander 1977 OFT 3 commander Resigned, 1979
MARITAL STATUS	Married, 4 children
SPACE EXPERIENCE	5 days 22 h 54 min 41 s

NAME	Terry Jonathan Hart
SPACE PERSON	139th (joint)
SERVICE RANK	Civilian
CURRENT STATUS	Manager, Bell Laboratories, Military and Government Systems Division
BIRTH DATE	27 October 1946
BIRTHPLACE	Pittsburg, Penn, USA
EXPERIENCE	Mechanical engineering degrees Electrical engineering degree US Air Force Reserve pilot Texas Air National Guard pilot Bell Telephone Laboratories
SPACE CAREER	NASA Group 8, August 1978 *Challenger* – STS 41C mission specialist, 6 April 1984 Resigned, 1984
MARITAL STATUS	Married, 1 child
SPACE EXPERIENCE	6 days 23 h 40 min 5 s

NAME	Henry Warren 'Hank' Hartsfield
SPACE PERSON	109th
SERVICE RANK	Col., US Air Force, retired
CURRENT STATUS	NASA astronaut commander
BIRTH DATE	21 November 1983
BIRTHPLACE	Birmingham, Alabama, USA
EXPERIENCE	Physics degree Astronautics degree US Air Force Test pilot Instructor
SPACE CAREER	USAF MOL Group 2, June 1966 NASA Group 7, August 1969 Apollo 16 support crew Skylab 2 support crew Skylab 3 support crew Skylab 4 support crew

	STS 2 back-up pilot STS 3 back-up pilot *Columbia* – STS 4 pilot, 27 June 1982 *Discovery* – STS 41D commander, 30 August 1984 STS 61A Spacelab D-1 commander
MARITAL STATUS	Married, 2 children
SPACE EXPERIENCE	13 days 2 h 6 min 3 s

Hartsfield. *NASA*

NAME	Frederick 'Rick' Hauck
SPACE PERSON	119th (joint)
SERVICE RANK	Capt., US Navy
CURRENT STATUS	NASA astronaut commander
BIRTH DATE	11 April 1941
BIRTHPLACE	Long Beach, California, USA
EXPERIENCE	Physics degree US Navy pilot Nuclear engineering degree Navy ROTC student Vietnam veteran 114 combat and support missions Executive officer, Attack Squadron
SPACE CAREER	NASA Group 8, August 1978 *Challenger* – STS 7 pilot, 18 June 1983 *Discovery* – STS 51A commander, 8 November 1984
MARITAL STATUS	Married, 2 children
SPACE EXPERIENCE	14 days 2 h 10 min 4 s

NAME	Steven Alan Hawley
SPACE PERSON	144th
SERVICE RANK	Civilian
CURRENT STATUS	NASA astronaut mission specialist
BIRTH DATE	12 December 1951
BIRTHPLACE	Ottowa, Kansas, USA
EXPERIENCE	Astronomy and physics degree Doctorate in astronomy
SPACE CAREER	NASA Group 8, August 1978 *Discovery* – STS 41D mission specialist, 30 August 1984 STS 61C mission specialist STS 61J mission specialist
MARITAL STATUS	Married (to NASA astronaut Sally Ride)
SPACE EXPERIENCE	6 days 0 h 56 min 4 s

NAME	Miroslaw Hermaszewski
SPACE PERSON	89th
SERVICE RANK	Col., Polish Air Force
CURRENT STATUS	Senior post, Polish Air Force
BIRTH DATE	15 September 1941
BIRTHPLACE	Lipniki, Poland
EXPERIENCE	Glider pilot Officer's Flying School Fighter pilot Commander of Squadron
SPACE CAREER	Intercosmos candidate, 1976 Soyuz 30 cosmonaut research, 27 June 1978 Retired
MARITAL STATUS	Unknown
SPACE EXPERIENCE	7 days 22 h 4 min

NAME:	Jeffrey Alan Hoffman
SPACE PERSON:	160th (joint)
SERVICE RANK:	Civilian
CURRENT STATUS:	NASA astronaut mission specialist
BIRTH DATE:	2 November 1944
BIRTHPLACE:	New York City, USA
EXPERIENCE:	Astronomy degree Doctorate in astrophysics
SPACE CAREER:	NASA Group 8, August 1978 *Discovery* – STS 51D mission specialist, 12 April 1985 STS 61E mission specialist
MARITAL STATUS:	Married, 2 children
SPACE EXPERIENCE:	6 days 23 hr 56 mins

NAME	James Benson Irwin
SPACE PERSON	54th (joint)
SERVICE RANK	Col., US Air Force, retired
CURRENT STATUS	Evangelist, Chairman High Flight Foundation
BIRTH DATE	17 March 1930
BIRTHPLACE	Pittsburgh, Penn, USA
EXPERIENCE	Naval sciences degree Aeronautical engineering degree Instrumentation engineering degree

	US Air Force pilot Aerospace research pilot
SPACE CAREER	NASA Group 5, April 1966 Apollo 10 support crew Apollo 12 back-up lunar module pilot Apollo 15 lunar module pilot, 26 July 1971 Apollo 17 back-up lunar module pilot, dropped Resigned, 1972
MARITAL STATUS	Married, 4 children
SPACE EXPERIENCE	12 days 7 h 11 min 53 s

Irwin. *NASA*

NAME	Alexander Sergeyevich Ivanchenkov
SPACE PERSON	88th
SERVICE RANK	Civilian
CURRENT STATUS	Cosmonaut
BIRTH DATE	28 September 1940
BIRTHPLACE	Ivanteyevka, USSR
EXPERIENCE	Moscow Aviation Institute Spacecraft designer
SPACE CAREER	Cosmonaut 1970 Soyuz 16 back-up flight engineer Soyuz 25 back-up flight engineer Original Soyuz 26 flight engineer Soyuz 26 back-up flight engineer Soyuz 27 back-up flight engineer Soyuz 29 flight engineer, 15 June 1978 Soyuz T-3 back-up flight engineer Soyuz T-6 flight engineer, 24 June 1982
MARITAL STATUS	Married, 1 child
SPACE EXPERIENCE	147 days 13 h 30 min

NAME	Georgi Ivan Ivanov
SPACE PERSON	92nd
SERVICE RANK	Col. Eng., Bulgarian Air Force
CURRENT STATUS	Unknown (non-active)
BIRTH DATE	2 July 1940
BIRTHPLACE	Lovech, Bulgaria
EXPERIENCE	Parachutist Glider pilot Air Force
SPACE CAREER	Intercosmos candidate, 1978 Soyuz 33, cosmonaut researcher, 10 April 1979 Retired
MARITAL STATUS	Married, 1 child
SPACE EXPERIENCE	1 day 23 h 1 min

NAME	Sigmund Jähn
SPACE PERSON	90th
SERVICE RANK	Col., East German Air Force
CURRENT STATUS	Unknown (non-active)
BIRTH DATE	13 February 1937
BIRTHPLACE	Rahtenbranz, East Germany
EXPERIENCE	Military Flight School Gagarin Air Force Academy
SPACE CAREER	Intercosmos candidate, 1976 Soyuz 31 cosmonaut researcher, 28 August 1978 Retired
MARITAL STATUS	Married, 2 children
SPACE EXPERIENCE	7 days 20 h 49 min

Ivanchenkov. *Novosti*

NAME	Joseph Peter Kerwin
SPACE PERSON	60th (joint)
SERVICE RANK	Capt., US Navy, MD
CURRENT STATUS	NASA astronaut mission specialist
BIRTH DATE	19 February 1932
BIRTHPLACE	Oak Hill, Illinois, USA
EXPERIENCE	Philosophy degree Medical degree Intern at Washington DC General Hospital Naval School of Aviation Medicine US Navy, pilot's wings Flight surgeon US Navy
SPACE CAREER	NASA Group 4, June 1965 Skylab 2 science pilot, 25 May 1973
MARITAL STATUS	Married, 3 children
SPACE EXPERIENCE	28 days 0 h 49 min 49 s

Kerwin. *NASA*

NAME	Yevgeni Vasilyevich Khrunov
SPACE PERSON	36th (joint)
SERVICE RANK	Col. Eng., Soviet Air Force
CURRENT STATUS	Official Star City
BIRTH DATE	10 September 1933
BIRTHPLACE	Prudy, USSR
EXPERIENCE	Agricultural college Air Force Aviator scientist degree
SPACE CAREER	Cosmonaut, March 1960 Voskhod 2 back-up pilot Soyuz 2 research engineer (cancelled) Soyuz 5 research engineer, 15 January 1969 Soyuz 38 back-up commander Soyuz 40 commander dropped from flight status Retired
MARITAL STATUS	Married, 1 child
SPACE EXPERIENCE	1 day 23 h 39 min

NAME	Leonid Denisovich Kizim
SPACE PERSON	98th (joint)
SERVICE RANK	Col., Soviet Air Force
CURRENT STATUS	Cosmonaut
BIRTH DATE	5 August 1941
BIRTHPLACE	Krasny Liman, USSR
EXPERIENCE	Military flight school 1963
SPACE CAREER	Cosmonaut, October 1965 Soyuz T-2 back-up commander Soyuz T-3 commander, 27 November 1980 Soyuz T-6 back-up commander Soyuz T-10-1 back-up commander Soyuz T-10 commander, 8 February 1984
MARITAL STATUS	Married, 2 children
SPACE EXPERIENCE	251 days 17 h 58 min

NAME	Pyotr Illyich Klimulk
SPACE PERSON	69th (joint)
SERVICE RANK	Maj. Gen., Soviet Air Force
CURRENT STATUS	Deputy commander, Cosmonaut Training Centre
BIRTH DATE	10 July 1942
BIRTHPLACE	Komarovka, Brest, USSR
EXPERIENCE	Military pilot
SPACE CAREER	Cosmonaut, October 1965 Soyuz 13 commander, 18 December 1973 Soyuz 17 back-up commander Soyuz 18-1 back-up commander Soyuz 18 commander, 24 May 1975 Soyuz 30 commander, 27 June 1978 Retired
MARITAL STATUS	Married, 1 child
SPACE EXPERIENCE	78 days 18 h 19 min

NAME	Vladimir Mikhailovich Komarov, deceased
SPACE PERSON	13th (joint)
SERVICE RANK	Col. Eng., Soviet Air Force Died 24 April 1967, Soyuz 1 crash
BIRTH DATE	16 March 1927
BIRTHPLACE	Moscow, USSR
EXPERIENCE	Pilot at 15 Graduated from four Air Force colleges
SPACE CAREER	Cosmonaut, March 1960 Vostok 4 back-up pilot Voskhod 1 commander, 12 October 1964 Soyuz 1 pilot, 23 April 1967, killed on landing
MARITAL STATUS	Married, 2 children
SPACE EXPERIENCE	2 days 3 h 5 min 3 s

NAME	Vladimir Vasilyevich Kovalyonok
SPACE PERSON	83rd (joint)
SERVICE RANK	Col., Soviet Air Force
CURRENT STATUS	Cosmonaut
BIRTH DATE	3 March 1942
BIRTHPLACE	Byeloye, Minsk, USSR
EXPERIENCE	Air Force School Transport aircraft pilot Paratroop instructor
SPACE CAREER	Cosmonaut 1967 Soyuz 18 back-up commander Soyuz 25 commander, 9 October 1977 Soyuz 26 back-up commander Soyuz 27 back-up commander Soyuz 29 commander, 15 June 1978 Soyuz T-3 back-up commander Soyuz T-4 commander, 12 March 1981
MARITAL STATUS	Married, 1 child
SPACE EXPERIENCE	216 days 9 h 12 min

Kovalyonok. *Novosti*

NAME	Valeri Nikolayevich Kubasov
SPACE PERSON	40th (joint)
SERVICE RANK	Civilian
CURRENT STATUS	Official, Mission Training, Star City
BIRTH DATE	7 January 1935
BIRTHPLACE	Vyasmki, USSR
EXPERIENCE	Aircraft engineer, 1958 Science degree
SPACE CAREER	Cosmonaut, August 1966 Soyuz 2 back-up flight engineer (cancelled) Soyuz 5 back-up flight engineer Soyuz 6 flight engineer, 11 October 1969 Soyuz 9 back-up flight engineer Salyut 1 flight engineer (reassigned) Soyuz 19/ASTP flight engineer, 15 July 1975 Soyuz 30 back-up commander Soyuz 36 commander, 26 May 1980 Retired
MARITAL STATUS	Unknown
SPACE EXPERIENCE	18 days 17 h 58 min 44 s

Kubasov. *Novosti*

NAME	Vasili Grigoryevich Lazarev
SPACE PERSON	64th (joint)
SERVICE RANK	Col., Soviet Air Force
CURRENT STATUS	Official at Star City
BIRTH DATE	23 February 1928
BIRTHPLACE	Altai region, Siberia, USSR
EXPERIENCE	Medical doctor Air Force pilot
SPACE CAREER	Cosmonaut, 1964 Voskhod 1 back-up doctor Soyuz 9 back-up commander Soyuz 12 commander, 27 September 1973 Soyuz 18-1 commander, 5 April 1975 Retired
MARITAL STATUS	Married, 1 child
SPACE EXPERIENCE	1 day 23 h 37 min 27 s

NAME	Valentin Vitalyevich Lebedev
SPACE PERSON	69th (joint)
SERVICE RANK	Civilian
CURRENT STATUS	Cosmonaut
BIRTH DATE	14 April 1942
BIRTHPLACE	Moscow, USSR
EXPERIENCE	Moscow Aviation Institute Spacecraft systems engineer
SPACE CAREER	Cosmonaut 1972 Soyuz 13 flight engineer, 18 December 1973 Soyuz 32 back-up flight engineer Soyuz 35 flight engineer, dropped due to injury Soyuz T5 flight engineer, 13 May 1982
MARITAL STATUS	Married, 1 child
SPACE EXPERIENCE	219 days 5 h

NAME	David Cornell Leestma
SPACE PERSON	149th (joint)
SERVICE RANK	Lt Cdr, US Navy
CURRENT STATUS	NASA astronaut mission specialist
BIRTH DATE	6 May 1949
BIRTHPLACE	Muskegan, Michigan, USA
EXPERIENCE	Aeronautical engineering degrees Test pilot
SPACE CAREER	NASA Group 9, May 1980 *Challenger* – STS 41G mission specialist, 5 October 1984 STS 61E mission specialist
MARITAL STATUS	Married, 1 child
SPACE EXPERIENCE	8 days 5 h 23 min 33 s

NAME	William Benjamin Lenoir
SPACE PERSON	112th (joint)
SERVICE RANK	Civilian
CURRENT STATUS	Management consultant
BIRTH DATE	14 March 1939
BIRTHPLACE	Miami, Florida, USA
EXPERIENCE	Electrical engineering degree Doctorate in electrical engineering
SPACE CAREER	NASA Group 6, August 1967 Skylab 3 back-up science pilot Skylab 4 back-up science pilot *Columbia* – STS 5 mission specialist, 11 November 1982 Resigned, 1984
MARITAL STATUS	Married, 2 children
SPACE EXPERIENCE	5 days 2 h 14 min 26 s

NAME	Alexei Archipovich Leonov
SPACE PERSON	16th (joint)
SERVICE RANK	Maj. Gen., Soviet Air Force
CURRENT STATUS	1st Deputy commander of Cosmonaut Training Centre
BIRTH DATE	20 May 1934
BIRTHPLACE	Altai region, Siberia, USSR
EXPERIENCE	Flying school Aeronautical engineering
SPACE CAREER	Cosmonaut, March 1960 Voskhod 2 pilot, 18 March 1965 Original Salyut 1 commander, reassigned to Soyuz 19-ASTP1 commander, 15 July 1975 Retired
MARITAL STATUS	Married, 2 children
SPACE EXPERIENCE	7 days 0 h 33 min 11 s

NAME	Byron Lichtenberg
SPACE PERSON	128th (joint)
SERVICE RANK	Civilian
CURRENT STATUS	Shuttle payload specialist
BIRTH DATE	19 February 1948
BIRTHPLACE	Stroudsburg, Penn, USA
EXPERIENCE	Electrical engineering degree US Air Force pilot Vietnam veteran Two DFCs and 11 air medals Mechanical engineering degree Doctorate in biomedical engineering Research staff member MIT
SPACE CAREER	Spacelab payload specialist candidate, 1977 *Columbia* – STS9/Spacelab payload specialist, 28 November 1983 STS 61K/EOM1 payload specialist
MARITAL STATUS	Married, 2 children
SPACE EXPERIENCE	10 days 7 h 47 min 23 s

Lenoir. *NASA*

NAME	Don Leslie Lind
SPACE PERSON	165th (joint)
SERVICE RANK	Commander US Navy Reserve
CURRENT STATUS	NASA astronaut mission specialist
BIRTH DATE	18 May 1930
BIRTHPLACE	Midvale, Utah, USA
EXPERIENCE	Science degree US Navy pilot Doctorate in high energy nuclear physics Space physicist
SPACE CAREER	NASA Group 5, April 1966 Skylab 3, back up science pilot Skylab 4, back up science pilot Skylab rescue stand-by pilot *Challenger* – STS 51B Spacelab 3 mission specialist, 29 April 1985
MARITAL STATUS	Married, 7 children
SPACE EXPERIENCE	7 days 0 h 8 min 50 s

NAME	Jack Robert Lousma
SPACE PERSON	62nd (joint)
SERVICE RANK	Col., USMC, retired
CURRENT STATUS	Space consultant
BIRTH DATE	29 February 1936
BIRTHPLACE	Grand Rapids, Michigan, USA
EXPERIENCE	Joined Marines Aeronautical engineering degree Pilot's wings
SPACE CAREER	NASA Group 5 1966 Apollo 9 support crew Apollo 10 support crew Apollo 13 support crew Original Apollo 17 back-up lunar module pilot Apollo 20 lunar module pilot (cancelled) Skylab 3 pilot, 28 July 1973 ASTP back-up docking module pilot OFT 3 pilot (rescheduled) *Columbia* – STS 3 commander, 22 March 1982 Resigned, 1983
MARITAL STATUS	Married, 4 children
SPACE EXPERIENCE	67 days 11 h 13 min 50 s

NAME	James Arthur Lovell Jr
SPACE PERSON	22nd (joint)
SERVICE RANK	Capt., US Navy, retired
CURRENT STATUS	Vice President, Centel Business Systems
BIRTH DATE	25 March 1928
BIRTHPLACE	Cleveland, Ohio, USA
EXPERIENCE	University of Wisconsin US Naval Academy US Navy pilot Test pilot Flight safety degree
SPACE CAREER	NASA Group 2 September 1962 Gemini 4 back-up pilot Gemini 7 pilot, 4 December 1965 Original Gemini 10 back-up command pilot Gemini 9 back-up command pilot Gemini 12 command pilot, 11 November 1966 Apollo 9 back-up command module pilot (renamed) Apollo 8 command module pilot, 21 December 1968 Apollo 11 back-up commander Apollo 13 commander, 11 April 1970 Resigned, 1973
MARITAL STATUS	Married, 4 children
SPACE EXPERIENCE	29 days 19 h 4 min 55 s

NAME	Vladimir Afanasevich Lyakhov
SPACE PERSON	91st
SERVICE RANK	Col., Soviet Air Force
CURRENT STATUS	Cosmonaut
BIRTH DATE	20 July 1941
BIRTHPLACE	Antratsit, USSR
EXPERIENCE	Air Force School Air Force
SPACE CAREER	Cosmonaut 1967 Soyuz 29 back-up commander Soyuz 32 commander, 25 February 1979 Soyuz 39 back-up commander Soyuz T-8 back-up commander Soyuz T-9 commander, 27 June 1983
MARITAL STATUS	Married, 2 children
SPACE EXPERIENCE	324 days 11 h 22 min

NAME	Oleg Grigoryevich Makarov
SPACE PERSON	64th (joint)
CURRENT STATUS	Unknown (non-active)
BIRTH DATE	16 January 1933
BIRTHPLACE	Kalinin, Nr Moscow, USSR
EXPERIENCE	Spacecraft designer
SPACE CAREER	Cosmonaut 1984 Voskhod 1 back-up scientist Soyuz 12 flight engineer, 27 September 1973 Soyuz 18-1 flight engineer, 5 April 1975 Soyuz 27 flight engineer, 10 January 1978 Soyuz T-2 back-up flight engineer Soyuz T-3 flight engineer, 27 November 1980 Retired
MARITAL STATUS	Married, 1 child
SPACE EXPERIENCE	20 days 17 h 44 min 27 s

NAME	Yuri Vasilyevich Malyshev
SPACE PERSON	95th
SERVICE RANK	Col., Soviet Air Force
CURRENT STATUS	Cosmonaut
BIRTH DATE	27 August 1941
BIRTHPLACE	Nikolayevsk, USSR
EXPERIENCE	Kharhov Higher Air Force School, 1963

SPACE CAREER	Cosmonaut 1967 Soyuz 22 back-up commander Soyuz T-2 commander, 5 June 1980 Soyuz T-6 commander (dropped) Soyuz T-11 commander, 3 April 1984
MARITAL STATUS	Married, 2 children
SPACE EXPERIENCE	11 days 19 h

NAME	Thomas Kenneth Mattingly
SPACE PERSON	56th (joint)
SERVICE RANK	Capt., US Navy
CURRENT STATUS	Commander, Electronic Systems, US Naval Command
BIRTH DATE	17 March 1936
BIRTHPLACE	Chicago, USA
EXPERIENCE	Aeronautics degree US Navy Pilot's wings Student at Aerospace Research Pilots School
SPACE CAREER	NASA Group 5, April 1966 Apollo 8 support crew Apollo 11 support crew Apollo 12 support crew Apollo 13 command module pilot, dropped two days before flight (suspected German Measles) Apollo 16 command module pilot, 16 April 1972 STS 2 back-up commander STS 3 back-up commander *Columbia* – STS 4 commander, 27 June 1982 *Discovery* – STS 51C commander, 24 January 1985 Resigned, 1985
MARITAL STATUS	Married, 1 child, divorced
SPACE EXPERIENCE	21 days 4 h 33 min 49 s

NAME	Jon Andrew McBride
SPACE PERSON	149th (joint)
SERVICE RANK	Commander, US Navy
CURRENT STATUS	NASA astronaut commander
BIRTH DATE	14 August 1943
BIRTHPLACE	Charleston, West Virginia, USA
EXPERIENCE	US Navy pilot Vietnam veteran 64 combat missions 3 air medals Aeronautical engineering degree
SPACE CAREER	NASA Group 8, August 1978 *Challenger* – STS 41G pilot, 5 October 1984 STS 61E commander
MARITAL STATUS	Married, 3 children
SPACE EXPERIENCE	8 days 5 h 23 min 33 s

NAME	Bruce McCandless
SPACE PERSON	132nd (joint)
SERVICE RANK	Capt., US Navy
CURRENT STATUS	NASA astronaut mission specialist
BIRTH DATE	8 June 1937
BIRTHPLACE	Boston, Mass., USA
EXPERIENCE	Naval sciences degree Electrical engineering degree US Navy pilot Skyray and Phantom pilot, USS *Enterprise*
SPACE CAREER	NASA Group 5, April 1966 Apollo 14 support crew Skylab 2 back-up pilot *Challenger* – STS 41B mission specialist, 3 February 1984 61J mission specialist
MARITAL STATUS	Married, 2 children
SPACE EXPERIENCE	7 days 23 h 15 min 54 s

Makarov. *Novosti*

NAME	James Alton McDivitt
SPACE PERSON	19th (joint)
SERVICE RANK	Brig. Gen., US Air Force, retired
CURRENT STATUS	Executive Vice President, Defense Electronics Operations
BIRTH DATE	10 June 1929
BIRTHPLACE	Chicago, Illinois, USA
EXPERIENCE	Plumber US Air Force 145 combat missions, Korea DFC and 5 air medals Aeronautical engineering Test pilot Aerospace research pilot
SPACE CAREER	NASA Group 2, September 1962 Gemini 4 command pilot, 3 June 1965 Original Apollo 1 back-up commander Apollo 2B commander (cancelled) Apollo 8 commander (mission renamed) Apollo 9 commander, 3 March 1969
MARITAL STATUS	Married, 4 children
SPACE EXPERIENCE	14 days 2 h 57 min 6 s

NAME	Ronald Erwin McNair
SPACE PERSON	132nd (joint)
SERVICE RANK	Civilian
CURRENT STATUS	NASA astronaut mission specialist
BIRTH DATE	21 October 1950
BIRTHPLACE	Lake City, South Carolina, USA
EXPERIENCE	Physics degree Doctorate in physics Doctorate in law Staff physicist, Hughes Research Labs
SPACE CAREER	NASA Group 8, August 1978 *Challenger* – STS 41B mission specialist, 3 February 1984 STS 51L mission specialist
MARITAL STATUS	Married, 1 child
SPACE EXPERIENCE	7 days 23 h 15 min 54 s

NAME	Arnaldo Tamayo Mendez
SPACE PERSON	97th
SERVICE RANK	Col., Cuban Air Force
CURRENT STATUS	Unknown (non-active)
BIRTH DATE	29 January 1942
BIRTHPLACE	Guantanamo, Cuba
EXPERIENCE	Association of Young Rebels Army pilot
SPACE CAREER	Intercosmos candidate, 1978 Soyuz 38 cosmonaut researcher, 18 September 1980 Retired
MARITAL STATUS	Married, 2 children
SPACE EXPERIENCE	7 days 20 h 43 min

McDivitt. *Rockwell*

NAME	Ulf Merbold
SPACE PERSON	128th (joint)
SERVICE RANK	Civilian
CURRENT STATUS	ESA astronaut payload specialist
BIRTH DATE	20 June 1941
BIRTHPLACE	Greiz, East Germany
EXPERIENCE	Defected to West Germany Doctorate in science Pilot's licence Physicist
SPACE CAREER	ESA payload specialist, 1977 *Columbia* – STS 9/Spacelab 1 payload specialist, 28 November 1983 STS 61A/Spacelab D1 back-up payload specialist
MARITAL STATUS	Married, 2 children
SPACE EXPERIENCE	10 days 7 h 47 min 23 s

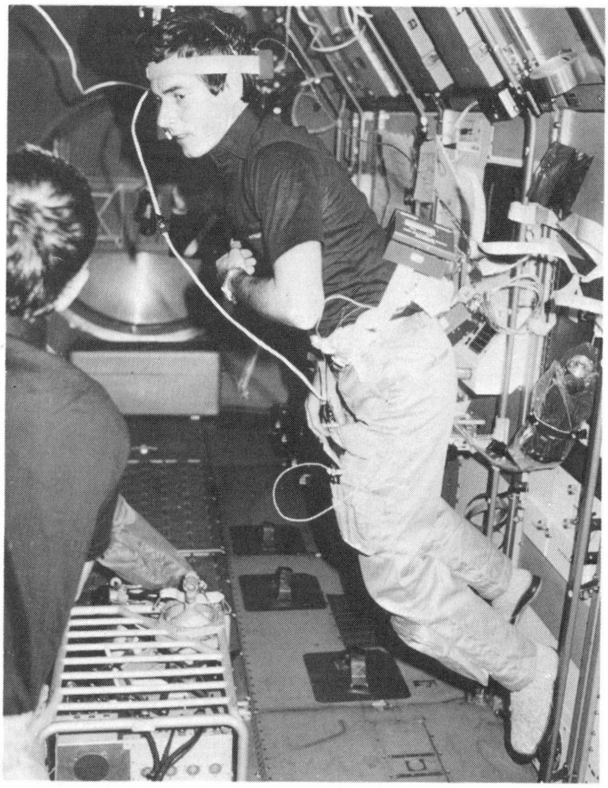

Merbold. *ESA*

NAME	Edgar Dean Mitchell
SPACE PERSON	49th (joint)
CURRENT STATUS	Chairman, Mitchell Communications Co
BIRTH DATE	17 September 1930
BIRTHPLACE	Hereford, Texas, USA
EXPERIENCE	Industrial management degree Aeronautical engineering degree Doctorate in aeronautics and astronautics US Navy Aerospace research pilot

SPACE CAREER	NASA Group 5, April 1966 Apollo 9 support crew Apollo 10 back-up lunar module pilot Apollo 14 lunar module pilot, 31 January 1971 Apollo 16 back-up lunar module pilot Resigned, 1972
MARITAL STATUS	Married, 2 children, divorced, remarried
SPACE EXPERIENCE	9 days 0 h 1 min 57 s

NAME	Richard Michael Mullane
SPACE PERSON	144th (joint)
SERVICE RANK	Lt Col., US Air Force
CURRENT STATUS	NASA astronaut mission specialist
BIRTH DATE	10 September 1945
BIRTHPLACE	Witchita Falls, Texas, USA
EXPERIENCE	Science degree, US Military Academy Vietnam veteran Weapons system operator 150 combat missions Aeronautical degree Air Force Institute of Technology
SPACE CAREER	NASA Group 8, August 1978 *Discovery* – STS 41D mission specialist, 30 August 1984 STS 62A mission specialist
MARITAL STATUS	Married, 3 children
SPACE EXPERIENCE	6 days 0 h 56 min 4 s

NAME	Franklin Story Musgrave
SPACE PERSON	115th (joint)
SERVICE RANK	Civilian, MD
CURRENT STATUS	NASA astronaut mission specialist
BIRTH DATE	19 August 1935
BIRTHPLACE	Boston, Mass., USA
EXPERIENCE	Mathematics/statistics degree Operations analysis and computer programming degree Chemistry degree Doctorate in medicine Physiology and biophysics degree
SPACE CAREER	NASA Group 6, August 1967 Skylab 2 back-up science pilot *Challenger* – STS6 mission specialist, 4 April 1983 STS 51F/Spacelab 2 mission specialist
MARITAL STATUS	Married, 5 children, divorced
SPACE EXPERIENCE	5 days 0 h 23 min 42 s

NAME	George Driver Nelson
SPACE PERSON	139th (joint)
SERVICE RANK	Civilian
CURRENT STATUS	NASA astronaut mission specialist
BIRTH DATE	13 July 1950
BIRTHPLACE	Charles City, Iowa, USA
EXPERIENCE	Physics degree Astronomy degree Doctorate in astronomy
SPACE CAREER	NASA Group 8, August 1978 *Challenger* – STS 41C mission specialist, 6 April 1984 STS 61C mission specialist
MARITAL STATUS	Married, 2 children
SPACE EXPERIENCE	6 days 23 h 40 min 5 s

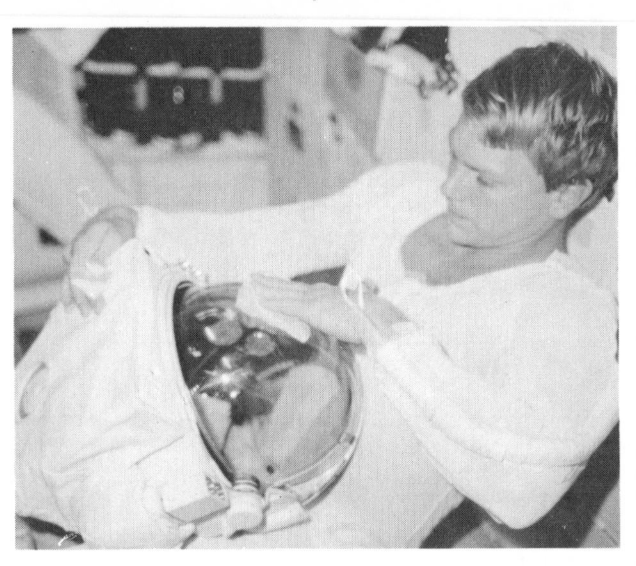

Nelson. *NASA*

NAME	Andrian Grigoryevich Nikolyev
SPACE PERSON	7th
SERVICE RANK	Maj. Gen., Soviet Air Force
CURRENT STATUS	Senior official, Mission Training, Star City
BIRTH DATE	5 September 1929
BIRTHPLACE	Shorshaly, Volga, USSR
EXPERIENCE	Medical student Lumberjack Air Force radio operator Air Force gunner Pilot's wings
SPACE CAREER	Cosmonaut, March 1960 Vostok 2 back-up pilot Vostok 3 pilot, 11 August 1961 Soyuz 2 back-up commander (cancelled) Soyuz 6, 7, 8 back-up commander Soyuz 9 commander, 1 June 1970 Retired
MARITAL STATUS	Married (to Valentina Tereshkova) 2 children, divorced
SPACE EXPERIENCE	21 days 15 h 20 min 50 s

NAME	Ellison Shoji Onizuka
SPACE PERSON	156th (joint)
SERVICE RANK	Maj., US Air Force
CURRENT STATUS	NASA astronaut mission specialist
BIRTH DATE	24 June 1946
BIRTHPLACE	Kealakekua, Hawaii, USA
EXPERIENCE	Science degrees US Air Force pilot Chief of Engineering Support Group, Edwards AFB

SPACE CAREER	NASA Group 8, August 1978 *Discovery* – STS 51C mission specialist, 24 January 1985 STS 51L mission specialist
MARITAL STATUS	Married, 3 children
SPACE EXPERIENCE	3 days 1 h 33 min 13 s

NAME	Robert Franklin Overmyer
SPACE PERSON	112th (joint)
SERVICE RANK	Col., US Marine Corps
CURRENT STATUS	NASA astronaut commander
BIRTH DATE	14 July 1936
BIRTHPLACE	Lorain, Ohio, USA
EXPERIENCE	Physics degree Aeronautics degree Aerospace research pilot
SPACE CAREER	USAF MOL Group 2, June 1966 NASA Group 7, August 1969 Apollo 17 support crew ASTP support crew OFT 4 pilot (reassigned) *Columbia* – STS 5 pilot, 11 November 1982 *Challenger* – STS 51B/Spacelab 3 commander, 29 November 1985
MARITAL STATUS	Married, 3 children
SPACE EXPERIENCE	12 days 2 h 23 min 16 s

Overmyer. *NASA*

NAME	Robert Allan Ridley Parker
SPACE PERSON	128th (joint)
SERVICE RANK	Civilian
CURRENT STATUS	NASA astronaut mission specialist
BIRTH DATE	14 December 1936
BIRTHPLACE	New York City, USA
EXPERIENCE	Astronomy, physics degree Doctorate in astronomy
SPACE CAREER	NASA Group 6, August 1967 Apollo 15 support crew Apollo 17 support crew and mission scientist *Columbia* – STS9/Spacelab 1 mission specialist, 28 November 1983 STS 61E mission specialist
MARITAL STATUS	Married, 2 children
SPACE EXPERIENCE	10 days 7 h 47 min 23 s

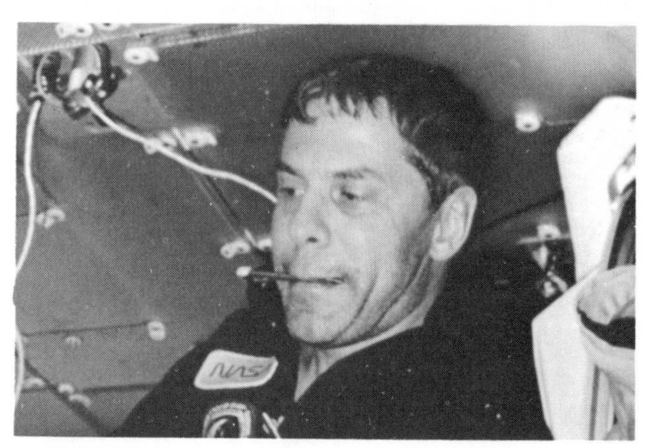

Parker. *NASA*

Nikolyev. *Novosti*

NAME	Viktor Ivanovich Patsayev, deceased
SPACE PERSON	52nd (joint)
SERVICE RANK	Civilian Died during Soyuz 11 descent, 29 June 1971
BIRTH DATE	19 June 1933
BIRTHPLACE	Aktyubinsk, USSR
EXPERIENCE	Master's degree from Industrial Institute Radio researcher Design engineer
SPACE CAREER	Cosmonaut, 1969 Soyuz 10 back-up research engineer Soyuz 11 research engineer, 6 June 1971, died during descent
MARITAL STATUS	Married, 2 children
SPACE EXPERIENCE	23 days 18 h 22 min

NAME	Gary Payton
SPACE PERSON	156th (joint)
SERVICE RANK	Maj., US Air Force
CURRENT STATUS	DoD manned spaceflight engineer
BIRTH DATE	1948
BIRTHPLACE	Rock Island, Illinois, USA
EXPERIENCE	Science degree, US Air Force Astronautical and aeronautical engineering degree US Air Force pilot Instructor Spacecraft test controller
SPACE CAREER	US Air Force manned spaceflight engineer, 1980 *Discovery* – STS 51C payload specialist, 24 January 1985
MARITAL STATUS	Married, 1 child
SPACE EXPERIENCE	3 days 1 h 33 min 13 s

NAME	Donald Herod Peterson
SPACE PERSON	115th (joint)
SERVICE RANK	Col., US Air Force, retired
CURRENT STATUS	Consultant
BIRTH DATE	22 October 1933
BIRTHPLACE	Winona, Mississippi, USA
EXPERIENCE	Science degree Nuclear engineering degree US Air Force pilot
SPACE CAREER	USAF MOL Group 3, June 1967 NASA Group 7, August 1969 Apollo 16 support crew *Challenger* – STS 6 mission specialist, 4 April 1983 Resigned, 1984
MARITAL STATUS	Married, 3 children
SPACE EXPERIENCE	5 days 0 h 23 min 42 s

NAME	William Reid Pogue
SPACE PERSON	66th (joint)
SERVICE RANK	Col., US Air Force, retired
CURRENT STATUS	Aerospace and energy consultant
BIRTH DATE	23 January 1930
BIRTHPLACE	Okemah, Oklahoma, USA
EXPERIENCE	Science degree Maths degree US Air Force pilot Korean War veteran Thunderbirds display team pilot
SPACE CAREER	NASA Group 5, April 1966 Apollo 1 support crew Apollo 7 support crew Apollo 11 support crew Apollo 13 support crew Apollo 14 support crew Original Apollo 16 back-up command module pilot Apollo 19 command module pilot (cancelled) Skylab 4 pilot, 16 November 1973 Resigned, 1975
MARITAL STATUS	Married, 3 children
SPACE EXPERIENCE	84 days 1 h 15 min 31 s

NAME	Leonid Ivanovich Popov
SPACE PERSON	93rd
SERVICE RANK	Col., Soviet Air Force
CURRENT STATUS	Cosmonaut
BIRTH DATE	31 August 1945
BIRTHPLACE	Alexandria, Ukraine, USSR
EXPERIENCE	Fitter electrician Air Force pilot school Commander, flight detachment, MIG 19s
SPACE CAREER	Cosmonaut, 1970 Soyuz 22 back-up commander Soyuz 32 back-up commander Soyuz 35 commander, 9 April 1980 Soyuz 40 commander, 15 May 1981 Soyuz T-7 commander, 19 August 1982
MARITAL STATUS	Married, 1 child
SPACE EXPERIENCE	200 days 14 h 42 min

NAME	Pavel Romanovich Popovich
SPACE PERSON	8th
SERVICE RANK	Maj. Gen., Soviet Air Force
CURRENT STATUS	Official at Star City
BIRTH DATE	5 October 1930
BIRTHPLACE	Uzin, near Kiev, USSR
EXPERIENCE	Shepherd Studied construction Air Force wings Order of Red Star for flying secret mission over Arctic
SPACE CAREER	Vostok 4 pilot, 12 August 1962 Soyuz 14 commander, 3 July 1974 Retired
MARITAL STATUS	Married, 2 children
SPACE EXPERIENCE	18 days 16 h 27 min

Popovich. *Novosti*

NAME	Dumitru Prunariu
SPACE PERSON	103rd
SERVICE RANK	Maj. Eng., Romanian Army Air Force
CURRENT STATUS	Unknown (non-active)
BIRTH DATE	27 September 1952
BIRTHPLACE	Brasov, Romania
EXPERIENCE	Aircraft factory worker Air Force Regiment of Romanian Army
SPACE CAREER	Intercosmos candidate, 1978 Soyuz 40 cosmonaut researcher, 15 May 1981 Retired
MARITAL STATUS	Married
SPACE EXPERIENCE	7 days 20 h 38 min

NAME	Vladimir Remek
SPACE PERSON	87th
SERVICE RANK	Col., Czech Air Force
CURRENT STATUS	Unknown (non-active)
BIRTH DATE	26 September 1948
BIRTHPLACE	Ceske-Budejovice, Czechoslovakia
EXPERIENCE	Air training school Gagarin Air Academy, USSR Czech Army pilot
SPACE CAREER	Intercosmos candidate, 1976 Soyuz 28 cosmonaut researcher, 2 March 1978 Retired
MARITAL STATUS	Married
SPACE EXPERIENCE	7 days 20 h 16 min

Remek. *Novosti*

NAME	Judith Arlene Resnik
SPACE PERSON	144th (joint)
SERVICE RANK	Civilian
CURRENT STATUS	NASA astronaut mission specialist
BIRTH DATE	5 April 1949
BIRTHPLACE	Akron, Ohio, USA
EXPERIENCE	Electrical engineering degree Doctorate in electrical engineering
SPACE CAREER	NASA Group 8, August 1978 *Discovery* – STS 41D, mission specialist, 30 August 1984 STS 51L mission specialist
MARITAL STATUS	Married, divorced
SPACE EXPERIENCE	6 days 0 h 56 min 4 s

NAME	Sally Kristen Ride
SPACE PERSON	119th (joint)
SERVICE RANK	Civilian
CURRENT STATUS	NASA astronaut mission specialist
BIRTH DATE	26 May 1951
BIRTHPLACE	Los Angeles, USA
EXPERIENCE	Physics degree English degree Science degree Doctorate in physics Research assistant, Stanford University
SPACE CAREER	NASA Group 8, August 1978 *Challenger* – STS 7 mission specialist, 18 June 1973 *Challenger* – STS 41G mission specialist, 5 October 1984
MARITAL STATUS	Married (to astronaut Steven Hawley)
SPACE EXPERIENCE	14 days 7 h 47 min 43 s

Ride. *NASA*

NAME	Yuri Viktorovich Romanenko
SPACE PERSON	85th
SERVICE RANK	Col., Soviet Air Force
CURRENT STATUS	Cosmonaut
BIRTH DATE	1 August 1944
BIRTHPLACE	Koltubanovsky, USSR
EXPERIENCE	Air Force college Flight instructor

SPACE CAREER	Cosmonaut, 1970 Soyuz 16 back-up commander Soyuz 25 back-up commander Soyuz 26 commander, 10 December 1977 Soyuz 33 back-up commander Soyuz 38 commander, 18 September 1980 Soyuz 40 back-up commander Soyuz T-7 back-up commander
MARITAL STATUS	Married, 1 child
SPACE EXPERIENCE	102 days 6 h 43 min

NAME	Stuart Allen Roosa
SPACE PERSON	49th (joint)
SERVICE RANK	Col., US Air Force, retired
CURRENT STATUS	President, Gulf Coast Coors, Inc.
BIRTH DATE	16 August 1933
BIRTHPLACE	Durango, Colorado, USA
EXPERIENCE	US Air Force Aeronautical engineering degree Test pilot Aerospace research pilot
SPACE CAREER	NASA Group 5, April 1966 Apollo 9 support crew Apollo 14 command module pilot, 31 January 1971 Apollo 16 back-up command module pilot Apollo 17 back-up command module pilot Resigned, 1976
MARITAL STATUS	Married, 4 children
SPACE EXPERIENCE	9 days 0 h 1 min 57 s

Roosa. *Gulf Coors*

NAME	Valeri Ilyich Rozhdestvensky
SPACE PERSON	80th (joint)
SERVICE RANK	Col. Eng., Soviet Air Force
CURRENT STATUS	Cosmonaut
BIRTH DATE	13 February 1939
BIRTHPLACE	Leningrad, USSR
EXPERIENCE	Naval engineer Diver-Cdr, Baltic Deep Sea Divers' Rescue Service Air Force
SPACE CAREER	Cosmonaut, 1965 Soyuz 21 back-up flight engineer Soyuz 23 flight engineer, 14 October 1976
MARITAL STATUS	Married, 1 child
SPACE EXPERIENCE	2 days 0 h 6 min

NAME	Nikolai Nikolayevich Ruchavishnikov
SPACE PERSON	51st
SERVICE RANK	Civilian
CURRENT STATUS	Head of Federation of Cosmonautics
BIRTH DATE	18 September 1932
BIRTHPLACE	Tomsk, Siberia, USSR
EXPERIENCE	Moscow Physics and Engineering Institute Spacecraft designer
SPACE CAREER	Cosmonaut, January 1967 Soyuz 6 and 7 back-up flight engineer Soyuz 10 research engineer, 23 April 1971 Soyuz 16 flight engineer, 2 December 1974 Soyuz 28 back-up commander Soyuz 33 commander, 10 April 1979 Soyuz T-11 flight engineer, dropped from flight status
MARITAL STATUS	Married, 1 child
SPACE EXPERIENCE	9 days 21 h 10 min

NAME	Valeri Viktorovich Ryumin
SPACE PERSON	83rd (joint)
SERVICE RANK	Civilian
CURRENT STATUS	Unknown (non-active)
BIRTH DATE	16 August 1939
BIRTHPLACE	Komsomolsk-on-Amur, Siberia, USSR
EXPERIENCE	Electronics engineer Spacecraft designer Designed Salyut 6
SPACE CAREER	Cosmonaut, 1973 Soyuz 25 flight engineer, 9 October 1977 Soyuz 29 back-up flight engineer Soyuz 32 flight engineer, 25 February 1979 Soyuz 35 flight engineer, 9 April 1980
MARITAL STATUS	Married, 2 children
SPACE EXPERIENCE	361 days 21 h 34 min

Inside a Soyuz trainer, April 1980. The marathon. The Soviets dominated in space in the 1970s. Crew after crew in Soyuz ferry vehicles docked with Salyut space stations for both short and long duration missions. The flight of Soyuz 35 was made by Leonid Popov and Valeri Ryumin, left. It lasted 184 days and made Ryumin the space record holder with nearly a year's space experience. (*Novosti*)

Space, 1980. Outlook on space. A view from Salyut showing a docked Soyuz T vehicle. (*Novosti*)

Moscow 1980. International spacemen. The Soviets flew non-Soviet cosmonauts on Intercosmos missions to the Salyut space station between 1978 and 1984. Seven of these cosmonauts were, left to right, Pham Tuan (Vietnam), Vladimir Remek (Czechoslovakia), Georgi Ivanov (Bulgaria), Miroslaw Hermaszewski (Poland), Bertalan Farkas (Hungary), Sigmund Jähn (East Germany) and Arnaldo Mendez (Cuba). (*Novosti*)

Earth orbit, 9 June 1980. Departure. The Soyuz T-2 ferry vehicle after departing from Salyut 6. (*Novosti*)

Inside Salyut 7, 20 August 1982. A lady visits. Twenty years after the first woman went into space, the Soviets launch Svetlana Savitskaya to visit the Salyut 7 space station in which cosmonauts Anatoli Berezovoi and Valentin Lebedev were in the middle of a 211 day endurance flight. (*Novosti*)

Near Arkalyk, USSR, 2 July 1982. Homecoming. The Soyuz T-6 capsule comes home with Frenchman Jean-Loup Chrétien and two Soviet companions on board. (*Novosti*)

Tyuratam, 27 September 1983. Abort. Soyuz T-10-1 was hauled away from its exploding rocket in the first launch pad abort in history. The cosmonauts were Gennadi Strekalov, right, and Vladimir Titov. (*Novosti*)

Ryumin. *Novosti*

NAME	Gennadi Vasilyevich Sarafanov
SPACE PERSON	72nd (joint)
SERVICE RANK	Col., Soviet Air Force
CURRENT STATUS	Official at Star Town
BIRTH DATE	1 January 1942
BIRTHPLACE	Saratov, USSR
EXPERIENCE	Military Aviation School Pilot
SPACE CAREER	Cosmonaut, October 1965 Soyuz 15 commander, 26 August 1974 Retired
MARITAL STATUS	Married, 2 children
SPACE CAREER	2 days 0 h 12 min

NAME	Viktor Savinykh
SPACE PERSON	100th
SERVICE RANK	Civilian
CURRENT STATUS	Cosmonaut
BIRTH DATE	7 March 1940
BIRTHPLACE	Berezhny village, Orichevesk, Kirov Oblast, USSR
EXPERIENCE	Railway engineer Aerial photographer Cartographer
SPACE CAREER	Cosmonaut, 1978 Soyuz T-3 back-up research engineer Soyuz T-4 flight engineer, 12 March 1981 Soyuz T-7 back flight engineer
MARITAL STATUS	Married, 1 child
SPACE EXPERIENCE	74 days 17 h 38 min

NAME	Svetlana Savitskaya
SPACE PERSON	110th (joint)
SERVICE RANK	Civilian
CURRENT STATUS	Cosmonaut
BIRTH DATE	8 August 1948
BIRTHPLACE	Moscow, USSR
EXPERIENCE	Pilot Flying instructor World aerobatic champion Test pilot Holder of 18 aviation world records
SPACE CAREER	Cosmonaut, 1980 Soyuz T-7 cosmonaut researcher, 18 August 1982 Soyuz T-12 flight engineer, 17 July 1984
MARITAL STATUS	Married
SPACE EXPERIENCE	19 days 17 h 6 min

Savitskaya. *Novosti*

NAME	Walter Marty Schirra Jr
SPACE PERSON	9th
SERVICE RANK	Capt., US Navy, retired
CURRENT STATUS	Vice President, Goodwin Coys, Colorado, USA
BIRTH DATE	12 March 1923
BIRTHPLACE	Hackensack, New Jersey, USA
EXPERIENCE	Engineering college Graduated US Naval Academy Pilot's wings, Pensacola Florida 71st Navy Fighter Squadron 90 combat missions, Korea Downed 2 MIG-15s, earned DFC and 2 air medals Test pilot

SPACE CAREER	NASA Group 1, 1959 MA-7 back-up pilot MA-8 pilot, 3 October 1962 Gemini 3 back-up command pilot Gemini 6 command pilot, 15 December 1965 Apollo 2A commander (cancelled) Apollo 1 back-up commander Apollo 7 commander, 11 October 1968 Resigned, 1969
MARITAL STATUS	Married, 2 children
SPACE EXPERIENCE	12 days 7 h 14 min 8 s

NAME	Harrison Hagen 'Jack' Schmitt
SPACE PERSON	58th (joint)
SERVICE RANK	Civilian
CURRENT STATUS	Space executive
BIRTH DATE	3 July 1935
BIRTHPLACE	Santa Rita, New Mexico, USA
EXPERIENCE	Science degree Doctorate in geology US Geological Survey's Astro-Geology Branch Instructed astronauts on field trips
SPACE CAREER	NASA Group 4, 1965 Apollo 15 back-up lunar module pilot Apollo 18 lunar module pilot (cancelled) Apollo 17 lunar module pilot, 7 December 1972 Resigned, 1975
MARITAL STATUS	Single
SPACE EXPERIENCE	12 days 13 h 51 min 59 s

Schmitt. *NASA*

NAME	Russell Louis 'Rusty' Schweickart
SPACE PERSON	39th
SERVICE RANK	Civilian
CURRENT STATUS	Chairman, California Energy Commission
BIRTH DATE	25 October 1935
BIRTHPLACE	Neptune, New Jersey, USA
EXPERIENCE	Aeronautical engineering degree US Air Force pilot Degrees in astronautics, aeronautics
SPACE CAREER	NASA Group 3, October 1963 Original Apollo 1 back-up pilot Apollo 2B lunar module pilot (rescheduled) Apollo 8 lunar module pilot (rescheduled) Apollo 9 lunar module pilot, 3 March 1969 Skylab 2 back-up commander Resigned, 1979
MARITAL STATUS	Married, 5 children
SPACE EXPERIENCE	10 days 1 h 0 min 54 s

NAME	Francis Richard 'Dick' Scobee
SPACE PERSON	139th (joint)
SERVICE RANK	Maj., US Air Force, retired
CURRENT STATUS	NASA astronaut commander
BIRTH DATE	19 May 1939
BIRTHPLACE	Cle Elum, Washington, USA
EXPERIENCE	US Air Force engine mechanic Aerospace engineering degree Pilot's wings Vietnam veteran Aerospace Research Pilot X-24B lifting body pilot Instructor pilot, SCA, shuttle carrier aircraft
SPACE CAREER	NASA Group 8, August 1978 *Challenger* – STS 41C pilot, 6 April 1984 STS 51L commander
MARITAL STATUS	Married, 2 children
SPACE EXPERIENCE	6 days 23 h 40 min 5 s

NAME	David Randolph Scott
SPACE PERSON	25th (joint)
SERVICE RANK	Col., US Air Force, retired
CURRENT STATUS	President, Scott Science and Technology
BIRTH DATE	6 June 1932
BIRTHPLACE	San Antonio, Texas, USA
EXPERIENCE	Science degree Aeronautics and astronautics Experimental test pilot Aerospace research pilot
SPACE CAREER	NASA Group 3, October 1963 Gemini 8 pilot, 16 March 1966 Original Apollo 1 back-up senior pilot Apollo 2B command module pilot (cancelled) Apollo 8 command module pilot (rescheduled)

Apollo 9 command module pilot, 3 March 1969
Apollo 12 back-up commander
Apollo 15 commander, 26 July 1971
Apollo 17 back-up commander (dropped)
Resigned, 1977

MARITAL STATUS Married, 2 children
SPACE EXPERIENCE 22 days 18 h 53 min 13 s

Scott. *NASA*

NAME Paul Scully-Power
SPACE PERSON 149th (joint)
SERVICE RANK Civilian
CURRENT STATUS Oceanographer, US Naval Underwater Systems
BIRTH DATE 1944
BIRTHPLACE Sydney, Australia
EXPERIENCE Applied mathematics degree
Scientific officer, RAN
Oceanographer
US citizen (1982)
SPACE CAREER Invited to fly as payload specialist, 1984
Challenger – STS 41G payload specialist, 5 October 1984
MARITAL STATUS Married, 6 children
SPACE EXPERIENCE 8 days 5 h 23 min 33 s

NAME: Margaret Rhea Seddon
SPACE PERSON: 160th (joint)
SERVICE RANK: Civilian
CURRENT STATUS: NASA astronaut mission specialist
BIRTHDATE: 8 November 1947
BIRTHDATE: Murfreesboro, Tennessee, USA
EXPERIENCE: Physiology degree
Doctor of medicine
Resident physician
SPACE CAREER: NASA Group 8, August 1978
Discovery – STS 51D mission specialist, 12 April 1985
STS 61D Spacelab 4/SL8-1 mission specialist
MARITAL STATUS: Married (to astronaut Robert Gibson), 1 child
SPACE EXPERIENCE: 6 days 23 h 56 min

NAME Alexander Alexandrov Serebrov
SPACE PERSON 110th (joint)
SERVICE RANK Civilian
CURRENT STATUS Cosmonaut
BIRTH DATE 15 February 1944
BIRTHPLACE Moscow, USSR
EXPERIENCE Spacecraft designer
SPACE CAREER Cosmonaut, 1978
Soyuz T-7 flight engineer, 19 August 1982
Soyuz T-8 flight engineer, 20 April 1983
MARITAL STATUS Unknown
SPACE EXPERIENCE 9 days 22 h 12 min

NAME Vitali Ivanovich Sevastyanov
SPACE PERSON 48th
SERVICE RANK Civilian
CURRENT STATUS Television presenter
BIRTH DATE 8 July 1935
BIRTHPLACE Krasnouralsk, USSR
EXPERIENCE Aviation Institute
Lecturer at Cosmonaut Training Centre
SPACE CAREER Cosmonaut, January 1967
Soyuz 6, 7 and 8 back-up flight engineer
Soyuz 9 flight engineer, 1 June 1970
Soyuz 17 back-up flight engineer
Soyuz 18-1 back-up flight engineer
Soyuz 18 flight engineer, 24 May 1975
Retired
MARITAL STATUS Married, 1 child
SPACE EXPERIENCE 80 days 16 h 18 min 50 s

NAME Rakesh Sharma
SPACE PERSON 138th
SERVICE RANK Squadron Leader, Indian Air Force
CURRENT STATUS Unknown (non-active)
BIRTH DATE 13 January 1949
BIRTHPLACE Patiala, India
EXPERIENCE Indian Air Force
SPACE CAREER Intercosmos candidate, 1982
Soyuz T-11 cosmonaut researcher, 3 April 1984
MARITAL STATUS Married, 2 children (1 died)
SPACE EXPERIENCE 7 days 21 h 41 min

NAME	Vladimir Alexandrovich Shatalov
SPACE PERSON	35th
SERVICE RANK	Lt Gen., Soviet Air Force
CURRENT STATUS	Dep. Commander, Cosmonaut Training Centre
BIRTH DATE	8 December 1927
BIRTHPLACE	Petropavlovsk, USSR
EXPERIENCE	Cadet pilot Instructor pilot Air Force Academy Squadron leader
SPACE CAREER	Cosmonaut, January 1963 Voskhod 3 back-up pilot (cancelled) Soyuz 3 back-up pilot Soyuz 4 commander, 14 January 1969 Soyuz 8 commander, 13 October 1969 Soyuz 10 commander, 23 April 1971 Retired
MARITAL STATUS	Married, 2 children
SPACE EXPERIENCE	9 days 21 h 57 min

NAME	Brewster Hopkinson Shaw Jr
SPACE PERSON	128th (joint)
SERVICE RANK	Lt Col., US Air Force
CURRENT STATUS	NASA astronaut commander
BIRTH DATE	16 May 1945
BIRTHPLACE	Cass City, Michigan, USA
EXPERIENCE	Engineering mechanics degree US Air Force Vietnam veteran
SPACE CAREER	NASA Group 8, August 1978 *Columbia* – STS 9 pilot, 28 November 1983 STS 61B commander
MARITAL STATUS	Married, 3 children
SPACE EXPERIENCE	10 days 7 h 47 min 23 s

NAME	Alan Bartlett Shepard Jr
SPACE PERSON	2nd
SERVICE RANK	Rear Admiral, US Navy, retired
CURRENT STATUS	President, Windward Coors Co, Texas
BIRTH DATE	18 November 1923
BIRTHPLACE	East Derry, New Hampshire, USA
EXPERIENCE	Science degree Navy pilot Test pilot Air Readiness Officer, Cdr in Chief, Atlantic Fleet
SPACE CAREER	NASA Group 1 astronaut, 1959 MR-3 pilot, 5 May 1961 MA-9 back-up pilot MA-10 pilot (cancelled) Original Gemini 3 commander, grounded – Menière's disease Flight status, 1969 Apollo 14 commander, 31 January 1971 Resigned, 1974
MARITAL STATUS	Married, 2 children
SPACE EXPERIENCE	9 days 0 h 17 min 25 s

NAME	Georgi Stepanovich Shonin
SPACE PERSON	40th (joint)
SERVICE RANK	Maj. Gen., Soviet Air Force
CURRENT STATUS	Head, cosmonaut shuttle team
BIRTH DATE	3 August 1935
BIRTHPLACE	Rovenki, Ukraine, USSR
EXPERIENCE	Aviation cadet Naval college Pilot's wings, Air Force
SPACE CAREER	Cosmonaut, March 1960 Voskhod 3 pilot (cancelled) Soyuz 5 back-up commander Soyuz 6 commander, 11 October 1969 Retired
MARITAL STATUS	Married, 2 children
SPACE EXPERIENCE	4 days 22 h 42 min

Shonin. *Novosti*

Shatalov. *Novosti*

NAME	Loren James Shriver
SPACE PERSON	156th (joint)
SERVICE RANK	Lt Col., US Air Force
CURRENT STATUS	NASA astronaut commander
BIRTH DATE	23 September 1944
BIRTHPLACE	Jefferson, Iowa, USA
EXPERIENCE	Science degree Astronautics degree US Air Force pilot Test pilot, Edwards AFB
SPACE CAREER	NASA Group 8, August 1978 *Discovery* – STS 51C pilot, 24 January 1985
MARITAL STATUS	Married, 4 children
SPACE EXPERIENCE	3 days 1 h 33 min 13 s

NAME	Donald Kent 'Deke' Slayton
SPACE PERSON	76th (joint)
SERVICE RANK	Maj., US Air Force, retired
CURRENT STATUS	President, Space Services Inc.
BIRTH DATE	1 March 1924
BIRTHPLACE	Sparta, Wisconsin, USA
EXPERIENCE	US Air Force wings B-25 bomber pilot, Second World War Aeronautical engineering degree Boeing test pilot US Air Force experimental test pilot
SPACE CAREER	NASA Group 1, April 1959 MA-7 pilot, dropped from flight status, March 1962. Restored to flight status, 1972 ASTP docking module pilot, 15 July 1975
MARITAL STATUS	Married, 1 child, divorced
SPACE EXPERIENCE	9 days 1 h 28 min 24 s

Slayton. *Space Services Inc.*

NAME	Vladimir Solovyov
SPACE PERSON	136th (joint)
SERVICE RANK	Civilian
CURRENT STATUS	Cosmonaut
BIRTH DATE	11 November 1946
BIRTHPLACE	Moscow, USSR
EXPERIENCE	Technical School graduate Space technologist
SPACE CAREER	Cosmonaut, 1978 Soyuz T-6 back-up flight engineer Soyuz T-9 back-up flight engineer Soyuz T-10 flight engineer, 8 February 1984
MARITAL STATUS	Married, 2 children
SPACE EXPERIENCE	236 days 22 h 50 min

NAME	Thomas Patten Stafford
SPACE PERSON	24th
SERVICE RANK	Lt Gen., US Air Force, retired
CURRENT STATUS	Consultant, Defense Technology
BIRTH DATE	17 September 1930
BIRTHPLACE	Weatherford, Oklahoma, USA
EXPERIENCE	US Naval Academy US Air Force Experimental test pilot Chief, Performance Branch Aerospace Research Pilot Group
SPACE CAREER	NASA Group 2, September 1962 Original Gemini 3 pilot Gemini 3 back-up pilot Gemini 6 pilot, 15 December 1965 Gemini 9 back-up commander (prime crew killed) Gemini 9 commander, 3 June 1966 Apollo 2A back-up senior pilot (cancelled) Apollo 2B back-up commander (cancelled) Apollo 7 back-up commander Apollo 10 commander, 18 May 1969 ASTP/Apollo 18 commander, 15 July 1975 Resigned, 1975
MARITAL STATUS	Married, 2 children
SPACE EXPERIENCE	21 days 3 h 44 min 31 s

NAME	Robert Lee Stewart
SPACE PERSON	132nd (joint)
SERVICE RANK	Col., US Army
CURRENT STATUS	NASA astronaut mission specialist
BIRTH DATE	13 August 1942
BIRTHPLACE	Washington DC, USA
EXPERIENCE	Mathematics degree US Army helicopter pilot Vietnam veteran Aerospace engineering degree
SPACE CAREER	NASA Group 8, August 1978 *Challenger* – STS 41B mission specialist, 3 February 1983 STS 51J mission specialist
MARITAL STATUS	Married, 2 children
SPACE EXPERIENCE	7 days 23 h 15 min 54 s

NAME	Gennadi Mikhailovich Strekalov
SPACE PERSON	98th (joint)
SERVICE RANK	Civilian
CURRENT STATUS	Cosmonaut
BIRTH DATE	28 October 1940
BIRTHPLACE	Mytischi, Moscow, USSR
EXPERIENCE	Baumann Technical School Spacecraft designer
SPACE CAREER	Cosmonaut, 1973 Soyuz 22 back-up flight engineer Soyuz 35 back-up flight engineer, dropped Soyuz T-3 research engineer, 27 November 1980 Soyuz T-4 back-up flight engineer Soyuz T-5 back-up flight engineer Soyuz T-8 flight engineer, 20 April 1983 Soyuz T-10-1 flight engineer, 27 September 1983 (aborted) Soyuz T-11 flight engineer, 3 April 1984
MARITAL STATUS	Married, 2 children
SPACE EXPERIENCE	22 days 17 h 9 min

Strekalov. *Novosti*

NAME	Kathryn Dwyer Sullivan
SPACE PERSON	149th (joint)
SERVICE RANK	Civilian
CURRENT STATUS	NASA astronaut mission specialist
BIRTH DATE	3 October 1951
BIRTHPLACE	Paterson, New Jersey, USA
EXPERIENCE	Earth sciences degree Doctorate in geology Systems engineer, remote sensing
SPACE CAREER	NASA Group 8, August 1978 *Challenger* – STS 41G mission specialist, 5 October 1984 61J mission specialist
MARITAL STATUS	Single
SPACE EXPERIENCE	8 days 5 h 23 min 33 s

NAME	John Leonard 'Jack' Swigert, deceased
SPACE PERSON	46th (joint)
SERVICE RANK	Civilian Died of cancer, 28 December 1982
BIRTH DATE	30 August 1931
BIRTHPLACE	Denver, Colorado, USA
EXPERIENCE	Mechanical engineering degree US Air Force pilot Korean War veteran Aerospace degree Test pilot, Pratt and Whitney
SPACE CAREER	NASA Group 5, April 1966 Apollo 1 support crew Apollo 7 support crew Apollo 11 support crew Apollo 13 back-up command module pilot Apollo 13 command module pilot, 11 April 1970 Leave of absence, 1973
MARITAL STATUS	Single
SPACE EXPERIENCE	5 days 22 h 54 min 41 s

Swigert. *NASA*

NAME	Valentina Vladirovna Tereshkova
SPACE PERSON	12th
SERVICE RANK	Col. Eng., Soviet Air Force, retired
CURRENT STATUS	Chairman, Soviet Women's Committee
BIRTH DATE	6 March 1937
BIRTHPLACE	Maslennikovo, Yaroslavl, USSR
EXPERIENCE	Cotton mill worker Amateur parachutist
SPACE CAREER	Cosmonaut, February 1962 Vostok 6, 16 June 1963 Retired
MARITAL STATUS	Married (to Andrian Nikolyev), 2 children, divorced
SPACE EXPERIENCE	2 days 22 h 50 min

NAME	Norman Earl Thagard
SPACE PERSON	119th (joint)
SERVICE RANK	Civilian
CURRENT STATUS	NASA astronaut mission specialist
BIRTH DATE	3 July 1943
BIRTHPLACE	Marianna, Florida, USA
EXPERIENCE	Engineering science degrees Doctorate of Medicine US Marine Corp pilot Vietnam veteran 163 combat missions Intern at South Carolina Hospital
SPACE CAREER	NASA Group 8, August 1978 *Challenger* – STS 7 mission specialist, 18 June 1983 *Challanger* – STS 51B/Spacelab 3 mission specialist, 29 April 1985 STS 61H mission specialist
MARITAL STATUS	Married, 2 children
SPACE EXPERIENCE	13 days 2 h 32 min

NAME	William Edgar Thornton
SPACE PERSON	124th (joint)
SERVICE RANK	Civilian, MD
CURRENT STATUS	NASA astronaut mission specialist
BIRTH DATE	14 April 1929
BIRTHPLACE	Faison, North Carolina, USA
EXPERIENCE	Physics degree Medical doctor
SPACE CAREER	NASA Group 6, August 1967 *Challenger* – STS8 mission specialist, 30 August 1983 *Challenger* – STS 51B/Spacelab 3 mission specialist, 29 April 1985
MARITAL STATUS	Married, 2 children
SPACE EXPERIENCE	13 days 1 h 17 min 30 s

Tereshkova. *Novosti*

NAME	Gherman Stepanovich Titov
SPACE PERSON	4th
SERVICE RANK	Lt Gen., Soviet Air Force
CURRENT STATUS	Soviet Ministry of Defence
BIRTH DATE	11 September 1935
BIRTHPLACE	Verkhneye Zhilino, Altai, USSR
EXPERIENCE	Cadet pilot Graduated with distinction from Air Force Pilots School
SPACE CAREER	Cosmonaut, March 1960 Vostok 1 back-up pilot Vostok 2 pilot, 6 August 1961 Retired
MARITAL STATUS	Married, 2 children
SPACE EXPERIENCE	1 day 1 h 18 min

NAME	Vladimir Georgyevich Titov
SPACE PERSON	118th (joint)
SERVICE RANK	Col., Soviet Air Force
CURRENT STATUS	Cosmonaut
BIRTH DATE	1 January 1947
BIRTHPLACE	Sretinsk, Chita region, USSR
EXPERIENCE	Test pilot
SPACE CAREER	Cosmonaut, 1976 Soyuz 35 back-up commander Soyuz T-4 back-up commander Soyuz T-5 back-up commander Soyuz T-8 commander, 20 April 1983 Soyuz T-10-1 commander, 27 September 1983 (aborted)
MARITAL STATUS	Married
SPACE EXPERIENCE	2 days 0 h 20 min

NAME	Richard Harrison Truly
SPACE PERSON	104th (joint)
SERVICE RANK	Commodore, US Navy
CURRENT STATUS	Commander, US Navy Space Command
BIRTH DATE	12 November 1937
BIRTHPLACE	Fayette, Mississippi, USA
EXPERIENCE	Aeronautical engineering degree US Navy Aerospace research pilot (ARP) ARP Instructor
SPACE CAREER	USAF MOL Group 1, December 1965 NASA Group 7, August 1969 Skylab 2 support crew Skylab 3 support crew Skylab 4 support crew ASTP support crew *Enterprise* – Shuttle ALT pilot, 1977 OFT pilot (redesignated) STS 1 back-up pilot *Columbia* – STS 2 pilot, 12 November 1981 *Challenger* – STS 8 commander, 30 August 1983 Resigned, 1983
MARITAL STATUS	Married, 3 children
SPACE EXPERIENCE	8 days 7 h 21 min 51 s

Truly. *Dara.*

NAME	Pham Tuan
SPACE PERSON	96th
SERVICE RANK	Lt Col., Vietnamese Air Force
CURRENT STATUS	Unknown (non-active)
BIRTH DATE	14 February 1947
BIRTHPLACE	Quoc Tuan, North Vietnam
EXPERIENCE	Air Force Vietnam war veteran Gagarin Air Force School, USSR
SPACE CAREER	Intercosmos candidate, 1979 Soyuz 37 cosmonaut researcher, 23 July 1980 Retired
MARITAL STATUS	Unknown
SPACE EXPERIENCE	7 days 20 h 42 min

NAME	Lodewijk van den Berg
SPACE PERSON	165th (joint)
SERVICE RANK	Civilian
CURRENT STATUS	Chemical engineer with EG and G Corporation
BIRTH DATE	24 March 1932
BIRTHPLACE	Sluiskil, the Netherlands (US Citizen)
EXPERIENCE	Chemical engineering degree Applied science degree Doctorate in applied science Expert in crystal growth
SPACE CAREER	Skylab 3 payload specialist candidate 1983 *Challenger* – STS 51B/Spacelab 3 payload specialist, 29 April 1985
MARITAL STATUS	Married, 2 children
SPACE EXPERIENCE	7 days 0 h 8 min 50 s

NAME	James Dougal Adrianus van Hoften
SPACE PERSON	139th (joint)
SERVICE RANK	Lt, US Navy, retired
CURRENT STATUS	NASA astronaut mission specialist
BIRTH DATE	11 June 1944
BIRTHPLACE	Fresno, California, USA
EXPERIENCE	US Navy aviator Civil engineering degree Hydraulic engineering degree Doctorate in fluid mechanics Assistant professor of civil engineering
SPACE CAREER	NASA Group 8, August 1978 *Challenger* – STS 41C mission specialist, 6 April 1984 STS 51I mission specialist
MARITAL STATUS	Married, 2 children
SPACE EXPERIENCE	6 days 23 h 40 min 5 s

NAME	Igor Volk
SPACE PERSON	143rd
SERVICE RANK	Col., Soviet Air Force, retired
CURRENT STATUS	Cosmonaut
BIRTH DATE	12 April 1937
BIRTHPLACE	Gottwald, USSR
EXPERIENCE	Military flight school Soviet Air Force pilot Test pilot Aviation Institute graduate
SPACE CAREER	Cosmonaut, 1978 Soyuz T-12 cosmonaut researcher, 17 July 1984
MARITAL STATUS	Unknown
SPACE EXPERIENCE	11 days 19 h 14 min

NAME	Vladislav Nikolayevich Volkov, deceased
SPACE PERSON	42nd (joint)
SERVICE RANK	Civilian Killed during Soyuz 11 descent, 29 June 1971
BIRTH DATE	23 November 1935
BIRTHPLACE	Moscow, USSR
EXPERIENCE	Aircraft designer Moscow Aviation Institute
SPACE CAREER	Cosmonaut, August 1966 Soyuz 7 flight engineer, 12 October 1969 Soyuz 10 back-up flight engineer Soyuz 11 flight engineer, 6 June 1971, killed during descent
MARITAL STATUS	Married, 1 child
SPACE EXPERIENCE	28 days 17 h 3 min

NAME	Boris Valentinovich Volynov
SPACE PERSON	36th (joint)
SERVICE RANK	Col., Soviet Air Force
CURRENT STATUS	Commander, Cosmonaut Group
BIRTH DATE	18 December 1934
BIRTHPLACE	Irkutsk, USSR

Volkov. *Novosti*

EXPERIENCE	Aviation college Soviet Air Force
SPACE CAREER	Cosmonaut, March 1960 Vostok 5 back-up pilot Voskhod 1 back-up commander Voskhod 3 commander (cancelled) Soyuz 5 commander, 15 January 1969 Soyuz 14 back-up commander Soyuz 15 back-up commander Soyuz 21 commander, 6 July 1976 Retired
MARITAL STATUS	Married, 2 children
SPACE EXPERIENCE	52 days 7 h 10 min

NAME	Charles David Walker
SPACE PERSON	144th (joint)
SERVICE RANK	Civilian
CURRENT STATUS	Manager, Electrophoresis Operations, McDonnell Douglas
BIRTH DATE	29 August 1948
BIRTHPLACE	Bedford, Indiana, USA
EXPERIENCE	Aeronautical and astronautical engineering degree Civil engineering technician Project engineer, Naval Sea Systems Shuttle test engineer, McDonnell Douglas Chief test engineer, Electrophoresis Operations, McDonnell Douglas
SPACE CAREER	Industry payload specialist, 1983 *Discovery* – STS 41D payload specialist, 30 August 1984 *Discovery* – STS 51D payload specialist, 12 April 1985 STS 61B payload specialist
MARITAL STATUS	Married
SPACE EXPERIENCE	13 days 0 h 52 min 4 s

NAME	David Mathieson Walker
SPACE PERSON	154th (joint)
SERVICE RANK	Commander, US Navy
CURRENT STATUS	NASA astronaut commander
BIRTH DATE	20 May 1944
BIRTHPLACE	Columbus, Georgia, USA
EXPERIENCE	US Navy Science degree Vietnam veteran
SPACE CAREER	NASA Group 8, August 1978 *Discovery* – STS 51A pilot, 8 November 1984
MARITAL STATUS	Married, 2 children
SPACE EXPERIENCE	7 days 23 h 45 min 54 s

NAME	Taylor G. Wang
SPACE PERSON	165th (joint)
SERVICE RANK	Civilian
CURRENT STATUS	Research scientist, Caltech, Jet Propulsion Laboratory
BIRTH DATE	16 June 1940
BIRTHPLACE	Shanghai, China (US citizen)
EXPERIENCE	Doctorate in physics Research scientist
SPACE CAREER	Spacelab 3 payload specialist candidate 1983 *Challenger* – STS 51B/Spacelab 3 payload specialist, 29 April 1985
MARITAL STATUS	Married, 2 children
SPACE EXPERIENCE	7 days 0 h 8 min 50 s

NAME	Paul Joseph Weitz
SPACE PERSON	60th (joint)
SERVICE RANK	Capt., US Navy, retired
CURRENT STATUS	NASA astronaut commander
BIRTH DATE	25 July 1932
BIRTHPLACE	Eire, Pennsylvania, USA
EXPERIENCE	Aeronautical engineering degree US Navy pilot Combat missions, Vietnam
SPACE CAREER	NASA Group 5, April 1966 Apollo 12 support crew Original Apollo 17 back-up command module pilot Apollo 20 command module pilot (cancelled) Skylab 2 pilot, 25 May 1973 *Challenger* – STS 6 commander, 4 April 1983
MARITAL STATUS	Married, 2 children
SPACE EXPERIENCE	33 days 1 h 13 min 31 s

NAME	Edward Higgins White 2nd, deceased
SPACE PERSON	19th (joint)
SERVICE RANK	Lt Col., US Air Force Killed in Apollo 1 fire, 27 January 1967
BIRTH DATE	14 November 1930
BIRTHPLACE	San Antonio, Texas, USA
EXPERIENCE	Science degree US Air Force Aeronautical engineering degree Test pilot

SPACE CAREER	NASA Group 2, September 1962 Gemini 4 pilot, 3 June 1965 Gemini 7 back-up command pilot Apollo 1 senior pilot, killed in spacecraft fire
MARITAL STATUS	Married, 2 children
SPACE EXPERIENCE	4 days 1 h 56 min 19 s

NAME	Donald Edward Williams
SPACE PERSON	160th (joint)
SERVICE RANK	Commander US Navy
CURRENT STATUS	NASA astronaut commander
BIRTHDATE	13 February 1942
BIRTHPLACE	Lafayette, Indiana, USA
EXPERIENCE	Mechanical engineering degree US Navy pilot Vietnam veteran
SPACE CAREER	NASA Group 8, August 1978 *Discovery* – STS 51D pilot, 12 April 1985
MARITAL STATUS	Married, 2 children
SPACE EXPERIENCE	6 days 23 h 56 min

NAME	Alfred Merrill Worden
SPACE PERSON	54th (joint)
SERVICE RANK	Lt Col., US Air Force, retired
CURRENT STATUS	President, Energy Management Consultant Co
BIRTH DATE	7 February 1932
BIRTHPLACE	Jackson, Michigan, USA
EXPERIENCE	Military Academy graduate US Air Force Empire Test Pilot School (UK) Aerospace research pilot Test pilot instructor
SPACE CAREER	NASA Group 5, April 1966 Apollo 9 support crew Apollo 12 back-up command module pilot Apollo 15 command module pilot, 26 July 1971 Apollo 17 back-up command module pilot, dropped from status. Resigned, 1975
MARITAL STATUS	Married, 2 children, divorced, remarried, 1 child
SPACE EXPERIENCE	12 days 7 h 11 min 53 s

NAME	Boris Borisovich Yegerov
SPACE PERSON	13th (joint)
SERVICE RANK	Medical Lt, Soviet Air Force, retired
CURRENT STATUS	Senior Medical Official, Manned Space Programme
BIRTH DATE	26 November 1937
BIRTHPLACE	Moscow, USSR
EXPERIENCE	Studied aviation medicine
SPACE CAREER	Cosmonaut, 1964 Vostok 7 pilot (cancelled) Voskhod 1 doctor, 12 October 1964 Retired
MARITAL STATUS	Married, 1 child
SPACE EXPERIENCE	1 day 0 h 17 min 3 s

NAME	Alexei Stanislovich Yeliseyev
SPACE PERSON	36th (joint)
SERVICE RANK	Civilian
CURRENT STATUS	Senior Official, Cosmonaut Training Centre
BIRTH DATE	13 July 1934
BIRTHPLACE	Zhizdra, USSR
EXPERIENCE	Master Technical Services Spacecraft designer
SPACE CAREER	Cosmonaut, August 1966 Soyuz 2 flight engineer (cancelled) Soyuz 5 flight engineer, 15 January 1969 Soyuz 8 flight engineer, 13 October 1969 Soyuz 10 flight engineer, 23 April 1971 Retired
MARITAL STATUS	Married, 1 child
SPACE EXPERIENCE	8 days 22 h 15 min

NAME	John Watts Young
SPACE PERSON	18th
SERVICE RANK	Capt., US Navy, retired
CURRENT STATUS	Chief of NASA astronauts
BIRTH DATE	24 September 1930
BIRTHPLACE	San Francisco, USA
EXPERIENCE	Aeronautical engineering US Navy Test pilot

Worden. *A. Worden*

SPACE CAREER	NASA Group 2 astronaut, September 1962 Gemini 3 pilot, 23 March 1965 Gemini 6 back-up pilot Gemini 10 commander, 18 July 1966 Apollo 2B back-up command module pilot (cancelled) Apollo 7 back-up senior pilot Apollo 10 command module pilot, 18 May 1969 Apollo 13 back-up commander Apollo 16 commander, 16 April 1972 Apollo 17 back-up commander OFT1 commander (redesignated) *Columbia* – STS 1 commander, 12 April 1981 *Columbia* – STS 9/Spacelab 1 commander, 28 November 1983
MARITAL STATUS	Married, 2 children, divorced, remarried
SPACE EXPERIENCE	34 days 19 h 42 min 13 s

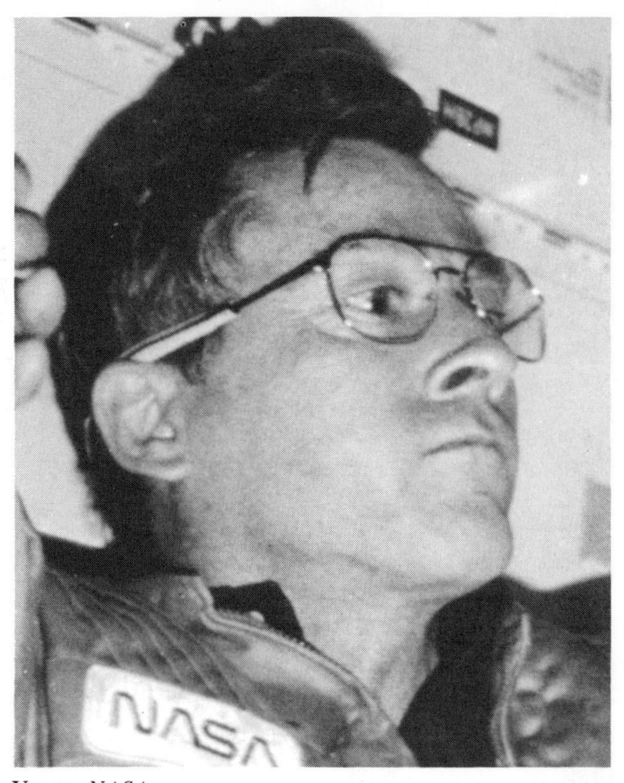

Young. *NASA*

NAME	Vitali Mikailovich Zholobov
SPACE PERSON	78th
SERVICE RANK	Col. Eng., Soviet Air Force
CURRENT STATUS	Unknown (non-active)
BIRTH DATE	18 June 1937
BIRTHPLACE	Zburyevka, USSR
EXPERIENCE	Oil and chemical industry Soviet Army Soviet Air Force
SPACE CAREER	Cosmonaut, January 1963 Soyuz 14 back-up flight engineer Soyuz 15 back-up flight engineer Soyuz 21 flight engineer, 6 July 1976 Retired
MARITAL STATUS	Married, 1 child
SPACE EXPERIENCE	49 days 6 h 24 min

NAME	Vyacheslav Dimitriyevich Zudov
SPACE PERSON	80th (joint)
SERVICE RANK	Lt Col., Soviet Air Force
CURRENT STATUS	Unknown (non-active)
BIRTH DATE	8 January 1942
BIRTHPLACE	Bor, near Gorky, USSR
EXPERIENCE	Graduated Military Aviation School Parachutist
SPACE CAREER	Cosmonaut, 1965 Back-up commander, Soyuz 21 Soyuz 23 commander, 14 October 1974 Retired
MARITAL STATUS	Married
SPACE EXPERIENCE	2 days 0 h 6 min

5

Manned Space Flight Tables

Manned Space Flight Duration Tables

Manned space flights in order of duration

Craft	days	h	min	s
Soyuz T-10	236	22	50	—
Soyuz T-5	211	8	5	—
Soyuz 35	185	20	12	—
Soyuz 32	175	—	36	—
Soyuz T-9	149	2	46	—
Soyuz 29	139	14	48	—
Soyuz 26	96	10	—	—
Skylab 4	84	1	15	31
Soyuz T-4	74	17	38	—
Soyuz 18	62	23	20	—
Skylab 3	59	11	9	4
Soyuz 21	49	6	24	—
Soyuz 17	29	13	20	—
Skylab 2	28	—	49	49
Soyuz 11	23	18	22	—
Soyuz 24	17	17	26	—
Soyuz 9	17	16	58	50
Soyuz 14	15	17	30	—
Gemini 7	13	18	35	1
Soyuz T-3	12	19	8	—
Apollo 17	12	13	51	59
Apollo 15	12	7	11	53
Soyuz T-12	11	19	14	—
Apollo 16	11	1	51	5
Apollo 7	10	20	9	3
STS 9	10	7	47	23
Apollo 12	10	4	36	25
Apollo 9	10	1	—	54
Apollo 18	9	1	28	—
Apollo 14	9	—	1	57
STS 41G	8	5	23	23
Apollo 11	8	3	18	35
STS 3	8	—	4	46
Apollo 10	8	—	3	23
STS 51A	7	23	45	54
STS 41B	7	23	15	54
Gemini 5	7	22	55	14
Soyuz T-6	7	22	42	—
Soyuz 30	7	22	4	—
Soyuz T-7	7	21	52	—
Soyuz 22	7	21	52	—
Soyuz T-11	7	21	41	—
Soyuz 13	7	20	55	—
Soyuz 31	7	20	49	—
Soyuz 36	7	20	46	—
Soyuz 39	7	20	43	—
Soyuz 38	7	20	43	—
Soyuz 37	7	20	42	—
Soyuz 40	7	20	38	—
Soyuz 28	7	20	16	—
STS 4	7	1	9	31
STS 51B	7	0	8	50
STS 51D	6	23	56	—
STS 41C	6	23	40	5
Apollo 8	6	3	—	42
STS 7	6	2	24	10
STS 8	6	1	8	40
STS 41D	6	—	56	4
Soyuz 27	5	22	59	—
Apollo 13	5	22	54	41
Soyuz 19	5	22	30	54
Soyuz 16	5	22	24	—
STS 5	5	2	14	26
STS 6	5	—	23	42
Vostok 5	4	23	6	—
Soyuz 8	4	22	51	—
Soyuz 6	4	22	42	—
Soyuz 7	4	22	41	—
Gemini 4	4	1	56	12
Soyuz 3	3	22	51	—
Gemini 12	3	22	34	31
Vostok 3	3	22	22	—
Soyuz T-2	3	22	19	—

Craft	days	h	min	s
STS 51C	3	1	33	13
Soyuz 5	3	—	54	—
Gemini 9	3	—	20	50
Soyuz 4	2	23	21	—
Gemini 11	2	23	17	8
Vostok 4	2	2	57	—
Vostok 6	2	22	50	—
Gemini 10	2	22	46	39
STS 1	2	6	20	52
STS 2	2	6	13	11
Soyuz 25	2	—	46	—
Soyuz T-8	2	—	20	—
Soyuz 15	2	—	12	—
Soyuz 23	2	—	6	—
Soyuz 10	1	23	45	—
Soyuz 12	1	23	16	—
Soyuz 33	1	23	1	—
MA-9	1	10	19	49
Soyuz 1	1	2	48	—
Voskhod 2	1	2	2	17
Gemini 6	1	1	51	54
Vostok 2	1	1	18	—
Voskhod 1	1	—	17	3
Gemini 8	—	10	41	26
MA-8	—	9	13	11
MA-7	—	4	56	5
MA-6	—	4	55	23
Gemini 3	—	4	52	51
Vostok 1	—	1	48	—
Soyuz 18-1	—	—	21	27
MR-4	—	—	15	37
MR-3	—	—	15	28

Nikolyev after Vostok 3. (*Novosti*)

Longest manned space flight

Soyuz T-10 (USSR) 236 days 22 h 50 min

Longest manned US space flight
Skylab 4 84 days 1 h 15 min 31 s

Shortest manned space flight

MR-3 (USA) 15 min 28 s

Shortest manned USSR space flight
Soyuz 18-1 21 min 27 s

Longest flights (by crew numbers)

(USSR has not flown crews of over 3 in number)

No. of crew	Flight	Country	days	h	min	s
1	Vostok 5	USSR	4	23	6	—
2	Soyuz T-5	USSR	211	8	5	—
3	Soyuz T-10	USSR	236	22	50	—
4	STS 5	USA	5	2	14	26
5	STS 51A	USA	7	23	45	54
6	STS 9	USA	10	7	47	23
7	STS 41G	USA	8	5	23	23

Longest US flights (by crew numbers)

No. of crew	Flight	days	h	min	s
1	MA-9	1	10	19	49
2	Gemini 7	13	18	35	1
3	Skylab 4	84	1	15	31

Shortest flights (by crew numbers)

No. of crew	Flight	Country	days	h	min	s
1	MR-3	USA	—	—	15	28
2	Soyuz 18-1	USSR	—	—	21	27
3	Voskhod 1	USSR	1	—	17	3
4	STS 6	USA	5	—	23	42
5	STS 51C	USA	3	1	33	13
6	STS 41D	USA	6	—	56	4
7	STS 51D	USA	6	23	56	—

Shortest USSR flight (1 crew)
Vostok 1 1 h 48 min

Shortest US flight (2 crew)
Gemini 3 4 h 52 min 51 s

Shortest US flight (3 crew)
Apollo 13 5 days 22 h 54 min 41 s

National manned spaceflight totals

	days	h	min	s
Total	2095	19	58	31
USSR	1661	21	1	31
USA	433	22	56	58

Other nations launched by either USSR or USA

	days	h	min	s
West Germany	10	7	47	23
Canada	8	5	23	33
France	7	22	42	—
Poland	7	22	4	—
India	7	21	41	—
East Germany	7	20	49	—
Hungary	7	20	46	—
Cuba	7	20	43	—
Mongolia	7	20	43	—
Vietnam	7	20	42	—
Romania	7	20	38	—
Czechoslovakia	7	20	16	—
Bulgaria	1	23	1	—

Man-hours in space

	days	h	min	s
USSR	3669	21	18	5
USA	1459	17	36	22
Total	5129	12	54	31

Manned space years

Year	No. of flights	US	USSR
1961	4	2	2
1962	5	3	2
1963	3	1	2
1964	1	—	1
1965	6	5	1
1966	5	5	—
1967	1	—	1
1968	3	2	1
1969	9	4	5
1970	2	1	1
1971	4	2	2
1972	2	2	—
1973	5	3	2
1974	3	—	3
1975	5	1	4
1976	3	—	3
1977	3	—	3
1978	5	—	5
1979	2	—	2
1980	6	—	6
1981	5	2	3
1982	6	3	3
1983	6	4	2
1984	8	5	3
1985	3	3	—
Total	105	48	57

Busiest year 1969
Quietest years 1964 and 1967

Years of greatest inequalities

Year	No. of flights	US	USSR
1966	5	5	—
1980	6	—	6

The launch of a Vostok. (*Novosti*)

People who have flown into space (Complete list of first flights)

No.	Name	Age	Flight	Country Placing
1	Yuri Gagarin	27	Vostok 1	USSR 1
2	Alan Shepard	38	*Freedom 7*	USA 1
3	Virgil Grissom	35	*Liberty Bell 7*	USA 2
4	Gherman Titov	25	Vostok 2	USSR 2
5	John Glenn	40	*Friendship 7*	USA 3
6	Scott Carpenter	37	*Aurora 7*	USA 4
7	Andrian Nikolyev	32	Vostok 3	USSR 3
8	Pavel Popovich	31	Vostok 4	USSR 4
9	Wally Schirra	39	*Sigma 7*	USA 5
10	Gordon Cooper	36	*Faith 7*	USA 6
11	Valeri Bykovsky	28	Vostok 5	USSR 5
12	Valentina Tereshkova	26	Vostok 6	USSR 6
13	Vladimir Komarov	37	Voskhod 1	USSR 7
13	Konstantin Feoktistov	38	Voskhod 1	USSR 7
13	Boris Yegerov	27	Voskhod 1	USSR 7
16	Pavel Belyayev	39	Voskhod 2	USSR 10
16	Alexei Leonov	30	Voskhod 2	USSR 10
18	John Young	35	Gemini 3	USA 7
19	James McDivitt	35	Gemini 4	USA 8
19	Edward White	34	Gemini 4	USA 8
21	Charles Conrad	35	Gemini 5	USA 10
22	Frank Borman	37	Gemini 7	USA 11
22	James Lovell	37	Gemini 7	USA 11
24	Thomas Stafford	35	Gemini 6	USA 13
25	Neil Armstrong	35	Gemini 8	USA 14
25	David Scott	34	Gemini 8	USA 14
27	Eugene Cernan	32	Gemini 9	USA 16
28	Michael Collins	36	Gemini 10	USA 17
29	Dick Gordon	37	Gemini 11	USA 18
30	Edwin Aldrin	36	Gemini 12	USA 19
31	Donn Eisele	38	Apollo 7	USA 20
31	Walter Cunningham	36	Apollo 7	USA 20
33	Georgi Beregovoi	47	Soyuz 3	USSR 12
34	William Anders	35	Apollo 8	USA 22
35	Vladimir Shatalov	42	Soyuz 4	USSR 13
36	Boris Volynov	34	Soyuz 5	USSR 14
36	Yevgeni Khrunov	34	Soyuz 5	USSR 14
36	Alexei Yeliseyev	35	Soyuz 5	USSR 14
39	Russel Schweickart	33	Apollo 9	USA 23
40	Georgi Shonin	34	Soyuz 6	USSR 17
40	Valeri Kubasov	34	Soyuz 6	USSR 17
42	Anatoli Filipchenko	41	Soyuz 7	USSR 19
42	Vladislav Volkov	33	Soyuz 7	USSR 19
42	Viktor Gorbatko	34	Soyuz 7	USSR 19
45	Alan Bean	37	Apollo 12	USA 24
46	Jack Swigert	39	Apollo 13	USA 25
46	Fred Haise	36	Apollo 13	USA 25
48	Vitali Sevastyanov	34	Soyuz 9	USSR 22
49	Stuart Roosa	37	Apollo 14	USA 27
49	Edgar Mitchell	40	Apollo 14	USA 27
51	Nikolai Ruchavishnikov	38	Soyuz 10	USSR 23
52	Georgi Dobrovolsky	43	Soyuz 11	USSR 24
52	Viktor Patsayev	38	Soyuz 11	USSR 24
54	Alfred Worden	39	Apollo 15	USA 29
54	James Irwin	41	Apollo 15	USA 29

No.	Name	Age	Flight	Country Placing
56	Thomas Mattingly	36	Apollo 16	USA 31
56	Charles Duke	36	Apollo 16	USA 31
58	Ronald Evans	39	Apollo 17	USA 33
58	Jack Schmitt	37	Apollo 17	USA 33
60	Joseph Kerwin	41	Skylab 2	USA 35
60	Paul Weitz	40	Skylab 2	USA 35
62	Owen Garriott	42	Skylab 3	USA 37
62	Jack Lousma	37	Skylab 3	USA 37
64	Vasili Lazarev	45	Soyuz 12	USSR 26
64	Oleg Makarov	40	Soyuz 12	USSR 26
66	Gerald Carr	41	Skylab 4	USA 39
66	Edward Gibson	37	Skylab 4	USA 39
66	William Pogue	43	Skylab 4	USA 39
69	Pyotr Klimuk	31	Soyuz 13	USSR 28
69	Valentin Lebedev	31	Soyuz 13	USSR 28
71	Yuri Artyukhin	44	Soyuz 14	USSR 30
72	Gennadi Sarafanov	32	Soyuz 15	USSR 31
72	Lev Demin	48	Soyuz 15	USSR 31
74	Alexei Gubarev	42	Soyuz 17	USSR 33
74	Georgi Grechko	42	Soyuz 17	USSR 33
76	Vance Brand	44	Apollo 18	USA 42
76	Deke Slayton	51	Apollo 18	USA 42
78	Vitali Zholobov	39	Soyuz 21	USSR 35
79	Vladimir Aksyonov	41	Soyuz 22	USSR 36
80	Vyacheslav Zudov	34	Soyuz 23	USSR 37
80	Valeri Rozhdestvensky	37	Soyuz 23	USSR 37
82	Yuri Glazkov	37	Soyuz 24	USSR 39
83	Vladimir Kovalyonok	35	Soyuz 25	USSR 40
83	Valeri Ryumin	38	Soyuz 25	USSR 40
85	Yuri Romanenko	33	Soyuz 26	USSR 42
86	Vladimir Dzhanibekov	35	Soyuz 27	USSR 43
87	Vladimir Remek	29	Soyuz 28	Czechoslovakia 1
88	Alexander Ivanchenkov	37	Soyuz 29	USSR 44
89	Miroslaw Hermaszewski	36	Soyuz 30	Poland 1
90	Sigmund Jähn	41	Soyuz 31	E. Germany 1
91	Vladimir Lyakhov	37	Soyuz 32	USSR 45
92	Georgi Ivanov	38	Soyuz 33	Bulgaria 1
93	Leonid Popov	34	Soyuz 35	USSR 46
94	Bertalan Farkas	30	Soyuz 36	Hungary 1
95	Yuri Malyshev	38	Soyuz T-2	USSR 47
96	Pham Tuan	33	Soyuz 37	Vietnam 1
97	Arnaldo Mendez	38	Soyuz 38	Cuba 1
98	Leonid Kizim	39	Soyuz T-3	USSR 48
98	Gennadi Strekalov	40	Soyuz T-3	USSR 48
100	Viktor Savinykh	41	Soyuz T-4	USSR 50
101	Jugderdemidyin Gurragcha	33	Soyuz 39	Mongolia 1
102	Robert Crippen	43	STS 1	USA 44
103	Dumitriu Prunariu	28	Soyuz 40	Romania 1
104	Joseph Engle	49	STS 2	USA 45
104	Richard Truly	44	STS 2	USA 45
106	Gordon Fullerton	45	STS 3	USA 47
107	Anatoli Berezovoi	40	Soyuz T-5	USSR 51
108	Jean-Loup Chrétien	44	Soyuz T-6	France 1
109	Henry Hartsfield	48	STS 4	USA 48
110	Alexander Serebrov	38	Soyuz T-7	USSR 52
110	Svetlana Savitskaya	34	Soyuz T-7	USSR 52
112	Robert Overmyer	46	STS 5	USA 49
112	Joseph Allen	43	STS 5	USA 49

No.	Name	Age	Flight	Country Placing
112	William Lenoir	45	STS 5	USA 49
115	Karol Bobko	45	STS 6	USA 52
115	Donald Peterson	49	STS 6	USA 52
115	Story Musgrave	47	STS 6	USA 52
118	Vladimir Titov	36	Soyuz T-8	USSR 54
119	Frederick Hauck	42	STS 7	USA 55
119	Sally Ride	32	STS 7	USA 55
119	John Fabian	44	STS 7	USA 55
119	Norman Thagard	39	STS 7	USA 55
123	Aleksander Aleksandrov	40	Soyuz T-9	USSR 55
124	Daniel Brandenstein	40	STS 8	USA 59
124	Guion Bluford	40	STS 8	USA 59
124	William Thornton	54	STS 8	USA 59
124	Dale Gardner	34	STS 8	USA 59
128	Brewster Shaw	38	STS 9	USA 63
128	Robert Parker	46	STS 9	USA 63
128	Byron Lichtenberg	35	STS 9	USA 63
128	Ulf Merbold	42	STS 9	W. Germany 1*
132	Robert Gibson	37	STS 41B	USA 66
132	Bruce McCandless	46	STS 41B	USA 66
132	Robert Stewart	41	STS 41B	USA 66
132	Ronald McNair	33	STS 41B	USA 66
136	Vladimir Solovyov	37	Soyuz T-10	USSR 56
136	Oleg Atkov	34	Soyuz T-10	USSR 56
138	Rakesh Sharma	35	Soyuz T-11	India 1
139	Francis Scobee	44	STS 41C	USA 70
139	George Nelson	33	STS 41C	USA 70
139	Terry Hart	37	STS 41C	USA 70
139	James van Hoften	39	STS 41C	USA 70
143	Igor Volk	47	Soyuz T-12	USSR 58
144	Michael Coats	38	STS 41D	USA 74
144	Judith Resnik	35	STS 41D	USA 74
144	Steven Hawley	32	STS 41D	USA 74
144	Richard Mullane	38	STS 41D	USA 74
144	Charles Walker	36	STS 41D	USA 74
149	Jon McBride	41	STS 41G	USA 79
149	Kathryn Sullivan	33	STS 41G	USA 79
149	David Leestma	35	STS 41G	USA 79
149	Paul Scully-Power	40	STS 41G	USA 79*
149	Marc Garneau	35	STS 41G	Canada 1
154	David Walker	40	STS 51A	USA 83
154	Anna Fisher	35	STS 51A	USA 83
156	Loren Shriver	40	STS 51C	USA 85
156	Elison Onizuka	38	STS 51C	USA 85
156	James Buchli	39	STS 51C	USA 85
156	Gary Payton	36	STS 51C	USA 85
160	Don Williams	42	STS 51D	USA 89
160	Rhea Seddon	37	STS 51D	USA 89
160	Jeff Hoffman	40	STS 51D	USA 89
160	Dave Giggs	45	STS 51D	USA 89
160	Jake Garn	52	STS 51D	USA 89
165	Fred Gregory	44	STS 51B	USA 94
165	Don Lind	54	STS 51B	USA 94
165	Lodewijk van den Berg	53	STS 51B	USA 94*
165	Taylor Wang	44	STS 51B	USA 94*

*Merbold born in East Germany, Scully-Power in Australia, van den Berg born in Holland, Wang in China.

People who have flown into space once

Age on first flight (men)

Oldest	Don Lind	54 (USA)
(*see* William Thornton, 56 under Two Flights)		
Youngest	Gherman Titov	25 (USSR)
Oldest Russian	Lev Demin	48
Youngest American	Eugene Cernan	32

Age on first flight (women)

Oldest	Rhea Seddon	37 (USA)
Youngest	Valentina Tereshkova	26 (USSR)
Oldest Russian	Svetlana Savitskaya	34
Youngest American	Sally Ride*	32

* Youngest American person in space.

Number of space people by nation

USA	97
USSR	58
Czechoslavkia	1
Poland	1
East Germany	1
Bulgaria	1
Hungary	1
Vietnam	1
Cuba	1
Mongolia	1
Romania	1
France	1
West Germany	1
India	1
Canada	1

People who have flown into space twice

No.	Name	Age	2nd flight	Country Placing
1	Virgil Grissom	39	Gemini 3	USA 1
2	Gordon Cooper (1st in orbit twice)	38	Gemini 5	USA 2
3	Walter Schirra	42	Gemini 6	USA 3
4	Thomas Stafford	35	Gemini 9	USA 4
5	John Young	36	Gemini 10	USA 5
6	Charles Conrad	36	Gemini 11	USA 6
7	James Lovell	38	Gemini 12	USA 7
8	Vladimir Komarov	40	Soyuz 1	USSR 1
9	Frank Borman	40	Apollo 8	USA 8
10	James McDivitt	39	Apollo 9	USA 9
10	David Scott	36	Apollo 9	USA 9
12	Eugene Cernan	35	Apollo 10	USA 11
13	Neil Armstrong	38	Apollo 11	USA 12
13	Michael Collins	38	Apollo 11	USA 12
13	Edwin Aldrin	39	Apollo 11	USA 12
16	Vladimir Shatalov	42	Soyuz 8	USSR 2
16	Alexei Yeliseyev	35	Soyuz 8	USSR 2
18	Richard Gordon	40	Apollo 12	USA 15
19	Andrian Nikolyev	40	Soyuz 9	USSR 4
20	Alan Shepard	47	Apollo 14	USA 16
21	Vladislav Volkov	35	Soyuz 11	USSR 5
22	Alan Bean	41	Skylab 3	USA 17
23	Pavel Popovich	43	Soyuz 14	USSR 6
24	Anatoli Filipchenko	46	Soyuz 16	USSR 7
24	Nikolai Ruchavishnikov	42	Soyuz 16	USSR 7
26	Vasili Lazarev	47	Soyuz 18-1	USSR 9
26	Oleg Makarov	42	Soyuz 18-1	USSR 9
28	Pyotr Klimuk	32	Soyuz 18	USSR 11
28	Vitali Sevastyanov	39	Soyuz 18	USSR 11
30	Alexei Leonov	41	Soyuz 19	USSR 13
30	Valeri Kubasov	40	Soyuz 19	USSR 13
32	Boris Volynov	41	Soyuz 21	USSR 15
33	Valeri Bykovsky	42	Soyuz 22	USSR 16
34	Viktor Gorbatko	42	Soyuz 24	USSR 17
35	Georgi Grechko	45	Soyuz 26	USSR 18
36	Alexei Gubarev	45	Soyuz 28	USSR 19
37	Vladimir Kovalyonok	36	Soyuz 29	USSR 20
38	Valeri Ryumin	39	Soyuz 32	USSR 21
39	Vladimir Aksyonov	45	Soyuz T-2	USSR 22
40	Yuri Romanenko	36	Soyuz 38	USSR 23
41	Vladimir Dzanhibekov	38	Soyuz 39	USSR 24
42	Leonid Popov	35	Soyuz 40	USSR 25
43	Jack Lousma	46	STS 3	USA 18
44	Valentin Lebedev	40	Soyuz T-5	USSR 26
45	Alexander Ivanchenkov	41	Soyuz T-6	USSR 27
46	Thomas Mattingly	46	STS 4	USA 19
47	Vance Brand	51	STS 5	USA 20
48	Paul Weitz	50	STS 6	USA 21
49	Alexander Serebrov	39	Soyuz T-8	USSR 28
49	Gennadi Strekalov	43	Soyuz T-8	USSR 28
51	Robert Crippen	45	STS 7	USA 22
52	Vladimir Lyakhov	42	Soyuz T-9	USSR 30
53	Richard Truly	45	STS 8	USA 23
54	Owen Garriott	53	STS 9	USA 24
55	Leonid Kizim	42	Soyuz T-10	USSR 31
56	Yuri Malyshev	42	Soyuz T-11	USSR 32
57	Svetlana Savitskaya	35	Soyuz T-12	USSR 33
58	Henry Hartsfield	50	STS 41D	USA 25
59	Sally Ride	33	STS 41G	USA 26
60	Frederick Hauck	43	STS 51A	USA 27
60	Joseph Allen	47	STS 51A	USA 27
60	Dale Gardner	36	STS 51A	USA 27

No	Name	Age	2nd flight	Country Placing
63	Karol Bobko	47	STS 51D	USA 30
63	Charlie Walker	36	STS 51D	USA 30
65	Robert Overmyer	48	STS 51B	USA 32
65	William Thornton	56	STS 51B	USA 32
65	Norman Thagard	41	STS 51B	USA 32

People who have flown into space twice

Age on second flight (men)

Oldest	William Thornton	56	(USA)
Youngest	Pyotr Klimuk	32	(USSR)
Oldest Russian	Vasili Lazarev	47	
Youngest American	Eugene Cernan	35	

Age on second flight (women)

Oldest	Svetlana Savitskaya	35	(USSR)
Youngest	Sally Ride	33	(USA)*

* Youngest American person in space twice

Total number of people who have flown into space twice

Americans	34
Russians	33

People who have flown into space three times

No.	Name	Age	3rd flight	Country Placing
1	Walter Schirra	45	Apollo 7	USA 1
2	James Lovell	40	Apollo 8	USA 2
3	Thomas Stafford	38	Apollo 10	USA 3
3	John Young	38	Apollo 10	USA 3
5	Charles Conrad	39	Apollo 12	USA 5
6	Vladimir Shatalov	43	Soyuz 10	USSR 1
6	Alexei Yeliseyev	36	Soyuz 10	USSR 1
8	David Scott	39	Apollo 15	USA 6
9	Eugene Cernan	38	Apollo 17	USA 7
10	Oleg Makarov	45	Soyuz 27	USSR 3
11	Pyotr Klimuk	35	Soyuz 30	USSR 4
12	Valeri Bykovsky	44	Soyuz 31	USSR 5
13	Nikolai Ruchavishnikov	46	Soyuz 33	USSR 6
14	Valeri Ryumin	40	Soyuz 35	USSR 7
15	Valeri Kubasov	45	Soyuz 36	USSR 8
16	Viktor Gorbatko	45	Soyuz 37	USSR 9
17	Vladimir Kovalyonok	39	Soyuz T-4	USSR 10
18	Vladimir Dzhanibekov	40	Soyuz T-6	USSR 11
19	Leonid Popov	36	Soyuz T-7	USSR 12
20	Vance Brand	52	STS 41B	USA 8
21	Gennadi Strekalov	43	Soyuz T-11	USSR 13
22	Robert Crippen	46	STS 41C	USA 9
23	Thomas Mattingly	48	STS 51C	USA 10

Age on third flight (men)

Oldest	Vance Brand	52 (USA)
Youngest	Pyotr Klimuk	35 (USSR)
Oldest Russian	Nikolai Ruchavishnikov	46
Youngest American	John Young	38

Total number of people who have flown into space three times

Americans	10
Russians	13

Apollo 15 crew departs. (*Author*)

People who have flown into space four times

No.	Name	Age	4th flight	Country Placing
1	James Lovell	42	Apollo 13	USA 1
2	John Young	41	Apollo 16	USA 2
3	Charles Conrad	42	Skylab 2	USA 3

No.	Name	Age	4th flight	Country Placing
4	Thomas Stafford	44	Apollo 18	USA 4
5	Oleg Makarov	47	Soyuz T-3	USSR 1
6	Vladimir Dzhanibekov	42	Soyuz T-12	USSR 2
7	Robert Crippen	47	STS 41G	USA 5

Age on fourth flight (men)

Oldest	Oleg Makarov	47 (USSR)
Youngest	John Young	42 (USA)
Oldest American	Robert Crippen	47
Youngest Russian	Vladimir Dzhanibekov	42

Total number of people who have flown into space four times

Americans	5
Russians	2

Fifth and sixth flights

John Young (USA) is the only person to have made five and six space flights on STS 1 (50) and STS 9 (53).

Total humans in space including the journeys of all those who flew more than once

Total	267	(9 women)
USSR	106	(3 women)
USA	148	(6 women)
Other nations	13	

Women who have flown into space once

No.	Name	Age	Flight	Country Placing
1	Valentina Tereshkova	26	Vostok 6	USSR 1
2	Svetlana Savitskaya	34	Soyuz T-7	USSR 2
3	Sally Ride	32	STS 7	USA 1
4	Judith Resnik	35	STS 41D	USA 2
5	Kathy Sullivan	33	STS 41G	USA 3
6	Anna Fisher	35	STS 51A	USA 4
7	Rhea Seddon	37	STS 51D	USA 5

Age on first flight

Youngest	Valentina Tereshkova	26 (USSR)
Oldest	Rhea Seddon	37 (USA)
Oldest Russian	Svetlana Savitskaya	34
Youngest American	Sally Ride	32

Women who have flown into space twice

Age on Second flight

No.	Name	Age	Flight	Country
1	Svetlana Savitskaya	35	Soyuz T–12	USSR
2	Sally Ride	33	STS 41G	USA

Over 50s 'space club'

	Flight ages	
'Deke' Slayton	51	USA
John Young	50, 53	USA
Paul Weitz	50	USA
William Thornton	54, 56	USA
Vance Brand	51, 52	USA
Hank Hartsfield	50	USA
'Jake' Garn	52	USA
Don Lind	54	USA
Lodewijk van den Berg	53	USA

The exhaust emitted by the space shuttle during launch is illustrated in this 35mm shot of 51D. (*Author*)

Space Experience Table

Name	Country	Time				No. of flights
		days	h	min	s	
Valeri Ryumin	USSR	361	21	34	—	3
Vladimir Lyakhov	USSR	324	11	22	—	2
Leonid Kizim	USSR	251	17	58	—	2
Vladimir Solovyov	USSR	236	22	50	—	1
Oleg Atkov	USSR	236	22	50	—	1
Valentin Lebedev	USSR	219	5	—	—	2
Vladimir Kovalyonok	USSR	216	9	12	—	3
Anatoli Berezovoi	USSR	211	8	5	—	1
Leonid Popov	USSR	200	14	42	—	3
Aleksander Aleksandrov	USSR	149	10	46	—	1
Alexander Ivanchenkov	USSR	147	13	30	—	2
Georgi Grechko	USSR	125	23	20	—	2
Yuri Romanenko	USSR	102	6	43	—	2
Gerald Carr	USA	84	1	15	31	1
Edward Gibson	USA	84	1	15	31	1
William Pogue	USA	84	1	15	31	1
Vitali Sevastyanov	USSR	80	16	18	50	2
Pyotr Klimuk	USSR	78	18	19	—	3
Viktor Savinykh	USSR	74	1	38	—	1
Owen Garriott	USA	69	18	56	27	2
Alan Bean	USA	69	15	45	29	2
Jack Lousma	USA	67	11	13	50	2
Boris Volynov	USSR	52	7	10	—	2
Vitali Zholobov	USSR	49	6	24	—	1
Charles Conrad	USA	49	3	38	36	4
Alexei Gubarev	USSR	37	7	36	—	2
John Young	USA	34	19	42	13	6
Vladimir Dzhanibekov	USSR	33	13	38	—	4
Paul Weitz	USA	33	1	13	31	2
Viktor Gorbatko	USSR	30	12	49	—	3
James Lovell	USA	29	19	4	55	4
Vladislav Volkov	USSR	28	17	3	—	2
Joseph Kerwin	USA	28	—	49	49	1
Georgi Dobrovolsky	USSR	23	18	22	—	1
Viktor Patsayev	USSR	23	18	22	—	1
Eugene Cernan	USA	23	14	16	12	3
Robert Crippen	USA	23	13	48	40	4
David Scott	USA	22	18	53	13	3
Gennadi Strekalov	USSR	22	17	9	—	3
Vance Brand	USA	22	2	58	44	3
Andrian Nikolyev	USSR	21	15	20	50	2
Ken Mattingly	USA	21	4	33	49	3
Thomas Stafford	USA	21	3	44	31	4
Valeri Bykovsky	USSR	20	17	47	—	3
Oleg Makarov	USSR	20	17	44	27	4
Frank Borman	USA	19	21	35	33	2
Svetlana Savitskaya	USSR	19	17	6	—	2
Valeri Kubasov	USSR	18	17	58	54	3
Pavel Popovich	USSR	18	16	27	—	2
Yuri Glazkov	USSR	17	17	26	—	1
Yuri Artyukhin	USSR	15	17	30	—	1
Sally Ride	USA	14	7	47	43	2
James McDivitt	USA	14	2	57	6	2
Rick Hauck	USA	14	2	10	4	2

Name	Country	Time				No. of flights
		days	h	min	s	
Dale Gardner	USA	14	—	54	35	2
Richard Gordon	USA	13	3	53	33	2
William Thornton	USA	13	2	32	0	2
Hank Hartsfield	USA	13	2	6	3	2
Joe Allen	USA	13	2	—	19	2
Norman Thagard	USA	13	1	17	30	2
Charles Walker	USA	13	0	52	4	2
Ronald Evans	USA	12	13	51	59	1
Jack Schmitt	USA	12	13	51	59	1
Wally Schirra	USA	12	7	14	8	3
Alfred Worden	USA	12	7	11	53	1
James Irwin	USA	12	7	11	53	1
Robert Overmyer	USA	12	2	23	16	2
Edwin Aldrin	USA	12	1	53	6	2
Karol Bobko	USA	12	0	19	42	2
Vladimir Aksyonov	USSR	11	20	11	—	2
Igor Volk	USSR	11	19	14	—	1
Yuri Malyshev	USSR	11	19	—	—	2
Michael Collins	USA	11	2	5	14	2
Charles Duke	USA	11	1	51	5	1
Anatoli Filipchenko	USSR	10	23	5	—	2
Donn Eisele	USA	10	20	9	3	1
Walt Cunningham	USA	10	20	9	3	1
Brewster Shaw	USA	10	7	47	23	1
Robert Parker	USA	10	7	47	23	1
Ulf Merbold	W.Germany	10	7	47	23	1
Byron Lichtenberg	USA	10	7	47	23	1
Russell Schweickart	USA	10	1	—	54	1
Alexander Serebrov	USSR	9	22	12	—	2
Vladimir Shatalov	USSR	9	21	57	—	3
Nikolai Ruchavishnikov	USSR	9	21	10	—	3
Gordon Cooper	USA	9	9	15	3	2
Deke Slayton	USA	9	1	28	24	1
Alan Shepard	USA	9	—	17	25	2
Stuart Roosa	USA	9	—	1	57	1
Edgar Mitchell	USA	9	—	1	57	1
Alexei Yeliseyev	USSR	8	22	15	—	3
Neil Armstrong	USA	8	14	—	1	2
Richard Truly	USA	8	7	21	51	2
Jon McBride	USA	8	5	23	33	1
Kathryn Sullivan	USA	8	5	23	33	1
David Leestma	USA	8	5	23	33	1
Marc Garneau	Canada	8	5	23	33	1
Paul Scully-Power	USA	8	5	23	33	1
Gordon Fullerton	USA	8	—	4	46	1
David Walker	USA	7	23	45	54	1
Anna Fisher	USA	7	23	45	54	1
Robert Gibson	USA	7	23	15	54	1
Bruce McCandless	USA	7	23	15	54	1
Ron McNair	USA	7	23	15	54	1
Robert Stewart	USA	7	23	15	54	1
Jean-Loup Chrétien	France	7	22	42	—	1
Miroslaw Hermaszewski	Poland	7	22	4	—	1
Rakesh Sharma	India	7	21	41	—	1
Sigmund Jähn	E.Germany	7	20	49	—	1
Bertalan Farkas	Hungary	7	20	46	—	1
Arnaldo Mendez	Cuba	7	20	43	—	1
Jugderdemidyin Gurragcha	Mongolia	7	20	43	—	1

Name	Country	Time days	h	min	s	No. of flights
Pham Tuan	Vietnam	7	20	42	—	1
Dumitru Prunariu	Romania	7	20	38	—	1
Vladimir Remek	Czech	7	20	16	—	1
Alexei Leonov	USSR	7	—	33	11	2
Frederick Gregory	USA	7	0	8	50	1
Don Lind	USA	7	0	8	50	1
Lodewijk van den Berg	USA	7	0	8	50	1
Taylor Wang	USA	7	0	8	50	1
Don Williams	USA	6	23	56	—	1
Rhea Seddon	USA	6	23	56	—	1
Jeff Hoffmann	USA	6	23	56	—	1
Dave Griggs	USA	6	23	56	—	1
'Jake' Garn	USA	6	23	56	—	1
Dick Scobee	USA	6	23	40	5	1
George Nelson	USA	6	23	40	5	1
James van Hoften	USA	6	23	40	5	1
Terry Hart	USA	6	23	40	5	1
William Anders	USA	6	3	—	42	1
John Fabian	USA	6	2	24	10	1
Daniel Brandenstein	USA	6	1	8	40	1
Guion Bluford	USA	6	1	8	40	1
Michael Coats	USA	6	—	56	4	1
Richard Mullane	USA	6	—	56	4	1
Steven Hawley	USA	6	—	56	4	1
Judith Resnik	USA	6	—	56	4	1
Jack Swigert	USA	5	22	54	41	1
Fred Haise	USA	5	22	54	41	1
William Lenoir	USA	5	2	14	26	1
Donald Peterson	USA	5	—	23	42	1
Story Musgrave	USA	5	—	23	42	1
Georgi Shonin	USSR	4	22	42	—	1
Edward White	USA	4	1	56	12	1
Georgi Beregovoi	USSR	3	22	51	—	1
Loren Shriver	USA	3	1	33	13	1
James Buchli	USA	3	1	33	13	1
Elison Onizuka	USA	3	1	33	13	1
Gary Payton	USA	3	1	33	13	1
Valentina Tereshkova	USSR	2	22	50	—	1
Joseph Engle	USA	2	6	13	11	1
Vladimir Komarov	USSR	2	3	5	3	2
Vladimir Titov	USSR	2	—	20	3	1
Gennadi Serafanov	USSR	2	—	12	—	1
Lev Demin	USSR	2	—	12	—	1
Vyacheslav Zudov	USSR	2	—	6	—	1
Yuri Rozhdestvensky	USSR	2	—	6	—	1
Yevgeni Khrunov	USSR	1	23	39	—	1
Vasili Lazarev	USSR	1	23	37	27	2
Georgi Ivanov	Bulgaria	1	23	1	—	1
Pavel Belyayev	USSR	1	2	2	17	1
Gherman Titov	USSR	1	1	18	—	1
Konstantin Feoktistov	USSR	1	—	17	3	1
Boris Yegerov	USSR	1	—	17	3	1
Gus Grissom	USA	—	5	8	26	2
Scott Carpenter	USA	—	4	56	5	1
John Glenn	USA	—	4	55	23	1
Yuri Gagarin	USSR	—	1	48	—	1

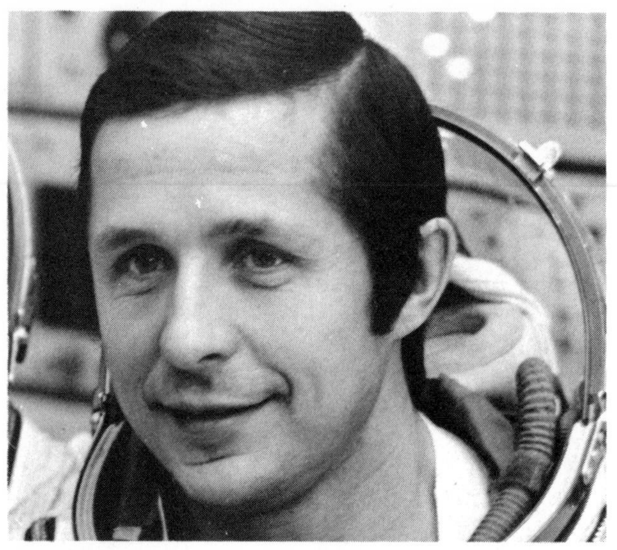

Savinykh. *Novosti*

Haise. (*Author*)

Most space experience

		days	h	min	s
Valeri Ryumin	USSR	361	21	34	—

Most space experience (US)

		days	h	min	s
Edward Gibson		84	1	15	31
Gerald Carr		84	1	15	31
Bill Pogue		84	1	15	31

Most space experience (Rest of the World)

		days	h	min	s
Ulf Merblod	W. Germany	10	7	47	23

Most female space experience

		days	h	min	s
Svetlana Savitskaya	USSR	19	17	6	—

Most female space experience (US)

		days	h	min	s
Sally Ride		14	7	47	43

Least space experience

		days	h	min	s
Yuri Gagarin	USSR	—	1	48	—

Least space experience (US)

		days	h	min	s
John Glenn		—	4	55	23

Least space experience (Rest of the World)

		days	h	min	s
Georgi Ivanov	Bulgaria	1	23	1	1

Least female space experience

		days	h	min	s
Valentina Tereshkova	USSR	2	22	50	1

Least female space experience (US)

		days	h	min	s
Judith Resnik		6	56	4	1

Most experience in space (by number of flights)

No. of flights		Country	days	h	min	s
1	Vladimir Solovyov	USSR	236	22	50	1
	Oleg Atkov	USSR	236	22	50	1
2	Leonid Kizim	USSR	251	17	58	—
3	Valeri Ryumin	USSR	361	21	34	—
4	Charles Conrad	USA	49	3	38	36
5	John Young	USA	24	1	54	50
6	John Young	USA	34	19	42	13
Most experience in space (US)						
1	Gerald Carr		84	1	15	31
	Edward Gibson		84	1	15	31
	Bill Pogue		84	1	15	31
2	Owen Garriott		69	18	56	27
3	Eugene Cernan		23	14	16	12
Most experience in space (USSR)						
4	Vladimir Dzhanibekov		33	13	38	—

Least experience in space (by number of flights)

No. of flights		Country	days	h	min	s
1	Yuri Gagarin	USSR	—	1	48	—
2	Gus Grissom	USA	—	5	8	26
3	Alexei Yeliseyev	USSR	8	22	15	—
4	Oleg Makarov	USSR	20	17	44	27
Least experience in space (US)						
1	John Glenn		—	4	55	23
3	Wally Schirra		12	7	14	8
4	Thomas Stafford		21	3	44	31
Least experience in space (USSR)						
2	Vasili Lazarev		1	23	37	27

Lunar Spacemen

One flight

	Name	Flight	
1	Frank Borman	Apollo 8	LO
1	James Lovell		
1	William Anders		
4	Tom Stafford	Apollo 10	LO
4	John Young		
4	Eugene Cernan		
7	Neil Armstrong	Apollo 11	LO
7	Michael Collins		
7	Buzz Aldrin		
10	Charles Conrad	Apollo 12	LO
10	Richard Gordon		
10	Alan Bean		
13	Jack Swigert*	Apollo 13	LF
13	Fred Haise		
15	Alan Shepard	Apollo 14	LO
15	Stuart Roosa		
15	Edgar Mitchell		
18	Dave Scott	Apollo 15	LO
18	Alfred Worden		
18	James Irwin		
21	Ken Mattingly	Apollo 16	LO
21	Charles Duke		
23	Ron Evans	Apollo 17	LO
23	Jack Schmitt		

(LO=Lunar orbit, LF=Lunar fly by) *Deceased.

Duke. *NASA*

Two flights

1	James Lovell	Apollo 13	LF
2	John Young	Apollo 16	LO
3	Eugene Cernan	Apollo 17	LO

Mitchell with the Apollo 14 Modular Equipment Transporter. (*NASA*)

Lunar walkers (in order of walk)

No.	Name	Flight	No. of EVAs	Total EVA time (depress to repress) days	h	min	s	Lunar stay time days	h	min	s
1	Neil Armstrong	Apollo 11	1	—	2	21	—	—	21	30	—
2	Buzz Aldrin	Apollo 11	1								
3	Charles Conrad	Apollo 12	2	—	7	45	—	1	7	31	—
4	Alan Bean	Apollo 12	2								
5	Alan Shepard	Apollo 14	2	—	9	22	—	1	9	31	—
6	Edgar Mitchell	Apollo 14	2								
7	David Scott	Apollo 15	3†	—	18	25	—*	2	18	55	—
8	James Irwin	Apollo 15	3								
9	John Young	Apollo 16	3†	—	20	14	—	2	23	14	—
10	Charles Duke	Apollo 16	3								
11	Eugene Cernan	Apollo 17	3†	—	22	5	—	3	2	59	—
12	Jack Schmitt	Apollo 17	3								
			TOTALS	3	8	22	—	12	11	40	

*Scott also performed stand-up EVA (33 min) from lunar module on the Moon soon after landing.
†Drove lunar roving vehicles.
See also EVA tables.

Command module lunar orbiters

Flight	Orbits	days	h	min	
Apollo 8	10	—	20	11	Borman/Lovell/Anders
Apollo 10	31	2	13	31	Young
Apollo 11	30	2	11	30	Collins
Apollo 12	45	3	14	56	Gordon
Apollo 14	34	2	18	39	Roosa
Apollo 15	74	6	1	18	Worden
Apollo 16	64	5	5	53	Mattingly*
Apollo 17	75	6	3	48	Evans

*Mattingly longest solo flight lunar orbit 3 days 9 h 28 min and longest US solo space flight.

Moon rock cargo

Flight	Weight (lb)
Apollo 11	48.5
Apollo 12	74.7
Apollo 14	98
Apollo 15	173
Apollo 16	213
Apollo 17	243
Total	850.2 lb

Leonov. *Novosti*

EVA Logs

Flight 1 (USSR 1) Voskhod 2
Alexei Leonov March 1965

Total depressurization time was 23 min 41 s. Leonov actually 'swam' in space for about 10 min. His 15-ft long tether contained telephone and telemetry cables. Air was provided by tanks contained in a backpack.

Flight 2 (USA 1) Gemini
Edward White June 1965

Total time outside the spacecraft was 21 min. Hatch open-to-close time was 36 min. White's 25-ft long tether provided him with oxygen; a ventilator control module, mounted on his chest, provided pressure and circulation in the suit. The VCM also had 9 min worth of emergency oxygen. White also used a hand-held manoeuvring unit, using oxygen gas.

Flight 3 (USA 2) Gemini 9
Eugene Cernan June 1966

Total hatch open-to-close time has been logged at 2 h 7 min. Cernan was equipped with an enlarged chest pack called Extravehicular Life Support System (ELSS) which also had a heat exchanger, to cool the ventilated air, and a 30 min emergency oxygen supply. The ELSS could not cope with Cernan's exertions and the astronaut's planned donning of an Astronaut Manoeuvring Unit (AMU) with independent oxygen supply had to be cancelled. Cernan also wore a new suit which was resistant to the hot gases that would have been expelled by the AMU. Tether length was 25 ft.

Flight 4 (USA 4) Gemini 10
Michael Collins July 1966

Collins performed a 49 min stand-up EVA, SEVA, with his head and chest protruding through Gemini's open hatch. He was attached by an umbilical. This SEVA was cut short because lithium hydroxide gas from Gemini's environmental control system seeped into the breathing air. During a 39 min EVA, curtailed because the command pilot needed to conserve manoeuvring fuel, Collins used the HHMU, this time powered by nitrogen from a separate tether. He also wore an ELSS and used a 50 ft long tether. There was also a 1 min 'equipment jettison' EVA, EJ.

Flight 5 (USA 4) Gemini 11
Richard Gordon September 1966

Gordon, using an ELSS, an HHMU and at the end of a 30 ft tether performed a 33 min EVA which was curtailed because, again, the ELSS could not cope with his efforts. The astronaut attached a tether to the docked Agena. There was also a 2 h 8 min SEVA and 2 min EJ.

Flight 6 (USA 5) Gemini 12
Edwin Aldrin November 1966

Aldrin performed two SEVAs lasting 2 h 29 min and 55 min and also a highly successful 2 h 6 min EVA, wearing an ELSS and using an HHMU.

Flight 7 (USSR 2) Soyuz 5-4
Yevgeni Khrunov, Alexei Yeleseyev
January 1969

During this EVA the cosmonauts were provided with air from backpacks and were attached by communications tethers. The EVA crew men transferred from Soyuz 5's orbital module to Soyuz 4. The exercise took 37 min.

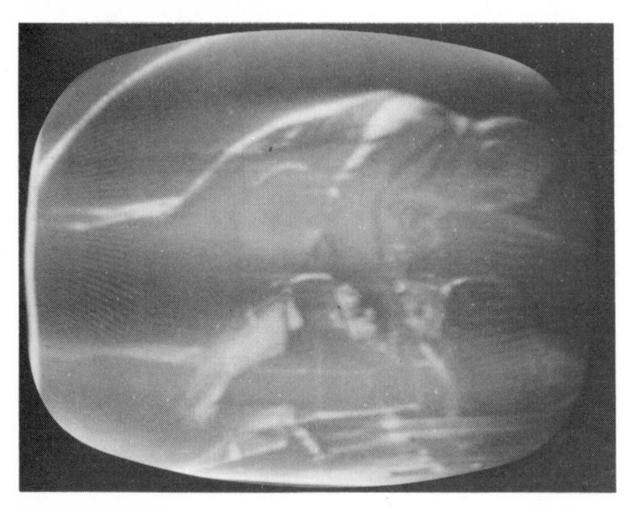

Soyuz 4/5 EVA. (*Novosti*)

Flight 8 (USA 6) Apollo 9
Russell Schweickart March 1969

During an EVA wearing the Apollo lunar spacesuit and Portable Life Support System backpack, during which he was to have transferred from the lunar module (*Spider*) to the command module (*Gumdrop*), Rusty Schweickart stood on the porch of the LM for about 56 min. He was still recovering from space sickness. Scott also performed a SEVA lasting about the same time.

Flight 9 (USA 7) Apollo 11
Neil Armstrong, Edwin Aldrin July 1969

One excursion onto the lunar surface lasted in total 2 h 40 min hatch open-to-close time. Armstrong was on the surface for 2 h 14 min and Aldrin for 1 h 33 min. There was also a brief EJ.

Cernan's Gemini 9 EVA. (*NASA*)

Flight 10 (USA 8) Apollo 12
Charles Conrad, Alan Bean
November 1969

There were two lunar walks, one lasting 3 h 56 min and the second 3 h 49 min (Conrad's times). Bean's times were 3 h 30 min and 3 h 20 min. Total EVA: 7 h 45 min.

Flight 11 (USA 9) Apollo 14
Alan Shepard, Edgar Mitchell
February 1971

Two LEVAs lasted a total of 9 h 24 min, hatch open-to-close times. The first lasted 4 h 49 min and the second 4 h 35 min. Shepard's times on the surface were 4 h 32 min and 4 h 22 min and Mitchell's were 4 h 20 min and 4 h 5 min.

Schweickart on the LM porch. (*NASA*)

Flight 12 (USA 10) Apollo 15
David Scott, James Irwin, Alfred Worden July/August 1971

There were five EVA exercises on this mission. Dave Scott performed a SEVA lasting 33 min, his head protruding out of the top of the lunar lander, soon after landing at Hadley Base. There were three Moon walks, involving the use of a lunar roving vehicle. Total walk times for each astronaut were Scott 17 h 36 min and Irwin 17 h 11 min. Scott's individual walk times were 6 h 14 min, 6 h 55 min and 4 h 27 min; Irwin's were 6 h 2 min, 6 h 50 min and 4 h 19 min. On the return journey Worden performed a Trans-Earth EVA (TEEVA). This lasted 38 min and, during it, Irwin was also outside but only on a SEVA.

Flight 13 (USA 11) Apollo 16
John Young, Charlie Duke, Ken Mattingly April 1972

A total of 20 h 15 min was spent outside Orion on the lunar surface during three LEVAs lasting 7 h 11 min, 7 h 23 min and 5 h 40 min. A lunar roving vehicle was used. Mattingly performed a 1 h 13 min TEEVA with Duke on SEVA.

Flight 14 (USA 12) Apollo 17
Eugene Cernan, Jack Schmitt, Ron Evans December 1972

The lunar roving vehicle was operated during three LEVA operations lasting 7 h 12 min, 7 h 37 min (the longest EVA on record) and 7 h 16 min. Total time was 22 h 4 min. Ron Evan's TEEVA lasted 1 h 6 min and during it Schmitt performed a SEVA.

Flight 15 (USA 13) Skylab 2
Paul Weitz, Joe Kerwin, Charles Conrad May/June 1973

Paul Weitz performed a 37 min SEVA soon after Apollo arrived at Skylab to try to prise out a jammed solar panel using a long pole. Conrad and Kerwin succeeded in deploying the panel during a 3 h 30 min space repair EVA. A second EVA later in the mission involved Conrad with Weitz. It lasted 1 h 44 min.

Flight 16 (USA 14) Skylab 3
Owen Garriott, Jack Lousma, Alan Bean July/September 1973

Three EVAs totalling 13 h 44 min were performed. Garriott and Lousma went EVA for 6 h 29 min, and again for 4 h 30 min while Bean and Garriott went outside for 2 h 45 min. Garriott's total is 13 h 44 min, Lousma's 10 h 59 min and Bean's 2 h 45 min.

Flight 17 (USA 15) Skylab 4
Edward Gibson, William Pogue, Gerald Carr November/February 1973–4

There were four EVAs on this flight lasting a total of 22 h 21 min. Gibson and Pogue were first in an EVA of 6 h 33 min. Carr was involved in the next three, one with Pogue then the last two with Gibson. These lasted 7 h 1 min, 3 h 28 min and 5 h 19 min. Carr's total is 15 h 48 min, Gibson's 15 h 17 min and Pogue 13 h 31 min.

Flight 18 (USSR 3) Soyuz 26
Georgi Grechko, Yuri Romanenko December/April 1977–8

During a single EVA lasting only 20 min, Grechko saved Romanenko's life when the latter cosmonaut on a SEVA, untethered, leaned too far out and became unofficially the first human satellite. Hatch open-to-close time was 1 h 28 min. Grechko wore a suit which incorporated a PLSS backpack.

Flight 19 (USSR 4) Soyuz 29
Vladimir Kovalyonok, Alexander Ivanchenkov July/October 1977

A single EVA lasted 2 h 5 min.

Flight 20 (USSR 5) Soyuz 32
Vladimir Lyakhov, Valeri Ryumin February/August 1979

One EVA lasting 1 h 23 min involved Ryumin's releasing a jammed antenna at the rear of Salyut 6.

Flight 21 (USSR 6) Soyuz T-5
Valentin Lebedev, Anatoli Berezovoi May/December 1982

There was one EVA on this flight lasting 2 h 33 min hatch open-to-close time.

The Skylab repair EVA. (*NASA*)

Flight 22 (USA 16) STS-6
Story Musgrave, Donald Peterson
April 1983

The first US space walk for nearly ten years lasted 4 h 17 min hatch open-to-close time and was conducted in the payload bay of the Shuttle *Challenger*. The astronauts were not tethered but used restraint harnesses. Total time outside was 3 h 54 min.

Flight 23 (USSR 7) Soyuz T-9
Vladimir Lyakhov, Aleksander Aleksandrov June/November 1983

Two EVAs lasted 2 h 50 min and 2 h 55 min. Lyakov's time is 5 h 45 min and Aleksandrov's the same.

Flight 24 (USA 17) STS-41B
Bruce McCandless, Robert Stewart
February 1984

EVA 1 lasted 5 h 35 min and during it both McCandless, first, and Stewart became human satellites wearing an MMU, which they also donned on EVA 2 lasting 6 h 2 min. At one point each of the pilots was 300 ft from the Shuttle. They did not operate MMUs in tandem. Total hatch open-to-close time was 12 h 12 min.

Flight 25 (USA 18) STS 41C
George Nelson, James van Hoften
April 1984

Wearing an MMU, Nelson flew to Solar Max and soft-docked. The EVA lasted 2 h 59 min and during it van Hoften remained in the payload bay. Solar Max was repaired during the longest EVA in Earth orbit on record, 7 h 7 min. At its end, Van Hoften tried out an MMU for a little spin. While total hatch open-to-close time was 10 h 6 min, NASA logs EVA time at 9 h 13 min.

Flight 26 (USSR 8) Soyuz T-10
Leonid Kizim, Vladimir Solovyov
February/October 1984

There were a record six EVAs on this mission, every one involving each cosmonaut, mainly to erect new solar panels on Salyut 7. Total time of 22 h 50 min was split as follows:

EVA	h min
1	3 45
2	4 56
3	2 45
4	2 45
5	3 05
6	5 approx

Soyuz T.10's Solovyov, Kizim and Atkov. (*Novosti*)

Flight 27 (USSR 9) Soyuz T-12
Vladimir Dhzanibekov, Svetlana Savitskaya July 1984

During this 3 h 35 min EVA, during which she operated welding equipment outside Salyut 7, Svetlana Savitskaya became the first woman to walk in space.

Flight 28 (USA 19) STS 41G
David Leestma, Kathy Sullivan
October 1984

This EVA lasted 3 h 29 min and during it the astronauts performed the first US space demonstration of fuel transfer.

Flight 29 (USA 20) STS 51A
Joe Allen, Dale Gardner
November 1984

The spectacular retrieval of two communications satellites featured in the two EVAs conducted on STS 51A. Allen used an MMU during the first EVA lasting 6 h and Gardner an MMU during the second 5 h 42 min EVA.

Flight 30 (USA 21) STS 51D
Jeff Hoffman, David Griggs
April 1985

This unscheduled and unrehearsed 3 h EVA was performed to fix jury-rigged 'fly swat' tools to the end of the RMS arm in a bid to flick a switch on a malfunctioning satellite, Syncom, that had been successfully deployed by the shuttle.

Note

It is impossible to be absolutely certain that these times are accurate. The main problem is that some official times are given as hatch open-to-close and some as actual total exposure. For example, Grechko was 20 min outside but the hatch open-to-close time was over 1 h.

Subsequently, the experience and record tables detailed below are restricted to those EVAs over 5 h and those spacemen with over 5 h EVA experience.

EVAs over 5 hours

Flight	(L)EVA	Country	Time h	Time min	Lunar/space walker
Apollo 17	LEVA 2	USA	7	37	Cernan, Schmitt
Apollo 16	LEVA 2	USA	7	23	Young, Duke
Apollo 17	LEVA 3	USA	7	16	Cernan, Schmitt
Apollo 17	LEVA 1	USA	7	12	Cernan, Schmitt
Apollo 16	LEVA 1	USA	7	11	Young, Duke
STS 41C	EVA 2	USA	7	07	Nelson, van Hoften
Skylab 4	EVA 2	USA	7	01	Carr, Pogue
Apollo 15	LEVA 2	USA	6	55	Scott, Irwin
Skylab 4	EVA 1	USA	6	33	Carr, Pogue
Skylab 3	EVA 1	USA	6	29	Garriott, Lousma
Apollo 15	LEVA 1	USA	6	14	Scott, Irwin
STS 41B	EVA 2	USA	6	02	McCandless, Stewart
STS 51A	EVA 1	USA	6*		Allen, Gardner
STS 51A	EVA 2	USA	5	42	Allen, Gardner
Apollo 16	EVA 3	USA	5	40	Young, Duke
STS 41B	EVA 1	USA	5	35	McCandless, Stewart
Skylab 4	EVA 4	USA	5	19	Carr, Gibson
Soyuz T-10	EVA 6	USSR	5*		Kizim, Solovyov

*Approximate times
LEVA = Lunar extravehicular activity (Moon walk)
EVA = Spacewalks in Earth orbit

EVA log: once

EVA	Name	Craft	Country Placing
1	Alexei Leonov	Voskhod 2	USSR 1
2	Edward White	Gemini 4	USA 1
3	Eugene Cernan	Gemini 9	USA 2
4	Mike Collins	Gemini 10	USA 3
5	Richard Gordon	Gemini 11	USA 4
6	Edwin Aldrin	Gemini 12	USA 5
7	Yevgeni Khrunov	Soyuz 5/4	USSR 2
8	Alexei Yeliseyev	Soyuz 5/4	USSR 3
9	Russell Schweickart	Apollo 9 *Spider*	USA 6
10	Neil Armstrong	Apollo 11 *Eagle*	USA 7
11	Charles Conrad	Apollo 12 *Intrepid*	USA 8

EVA	Name	Craft	Country Placing
12	Alan Bean	Apollo 12 *Intrepid*	*USA 9*
13	Alan Shepard	Apollo 14 *Antares*	USA 10
14	Edgar Mitchell	Apollo 14 *Antares*	USA 11
15	David Scott	Apollo 15 *Falcon*	USA 12
16	James Irwin	Apollo 15 *Falcon*	USA 13
17	Alfred Worden	Apollo 15 *Endeavour*	USA 14
18	John Young	Apollo 16 *Orion*	USA 15
19	Charles Duke	Apollo 16 *Orion*	USA 16
20	Ken Mattingly	Apollo 16 *Casper*	USA 17
21	Jack Schmitt	Apollo 17 *Challenger*	USA 18
22	Ron Evans	Apollo 17 *America*	USA 19
23	Paul Weitz	Skylab 2	USA 20
24	Joe Kerwin	Skylab 2	USA 21
25	Owen Garriott	Skylab 3	USA 22
26	Jack Lousma	Skylab 3	USA 33
27	Edward Gibson	Skylab 4	USA 24
28	Bill Pogue	Skylab 4	USA 25
29	Gerald Carr	Skylab 4	USA 26
30	Georgi Grechko	Soyuz 36 Salyut 6	USSR 4
31	Yuri Romanenko	Soyuz 36 Salyut 6	USSR 5*
32	Vladimir Kovalyonok	Soyuz 29 Salyut 6	USSR 6
33	Alexander Ivanchenkov	Soyuz 29 Salyut 6	USSR 7
34	Vladimir Lyakhov	Soyuz 32 Salyut 6	USSR 8
35	Vladimir Ryumin	Soyuz 32 Salyut 6	USSR 9
36	Anatoli Berezovoi	Soyuz T-5 Salyut 7	USSR 10
37	Valentin Lebedev	Soyuz T-5 Salyut 7	USSR 11
38	Story Musgrave	STS 6 *Challenger*	USA 27
39	Don Peterson	STS 6 *Challenger*	USA 28
40	Aleksander Aleksandrov	Soyuz T-9 Salyut 7	USSR 12
41	Bruce McCandless	STS 41B *Challenger* MMU	USA 29
42	Robert Stewart	STS 41B *Challenger* MMU	USA 30
43	George Nelson	STS 41C *Challenger* MMU	USA 31
44	James van Hoften	STS 41C *Challenger*	USA 32
45	Leonid Kizim	Soyuz T-10 Salyut 7	USSR 13
46	Vladimir Solovyov	Soyuz T-10 Salyut 7	USSR 14
47	Vladimir Dzhanibekov	Soyuz T-12 Salyut 7	USSR 15
48	Svetlana Savitskaya†	Soyuz T-12 Salyut 7	USSR 16
49	David Leestma	STS 41G *Challenger*	USA 33
50	Kathryn Sullivan†	STS 41G *Challenger*	USA 34
51	Joe Allen	STS 51A *Discovery* MMU	USA 35
52	Dale Gardner	STS 51A *Discovery*	USA 36
53	Jeff Hoffman	STS 51D *Discovery*	USA 37
54	Dave Griggs	STS 51D *Discovery*	USA 38

*Romanenko floated out of the airlock untethered and was rescued with an instinctive grab by Grechko, so in effect he made the first untethered spacewalk. McCandless made the first official one.

†Female.

EVA log: twice

EVA	Name	Craft	Country Placing
1	Edwin Aldrin	Apollo 11 *Eagle*	USA 1
2	Charles Conrad	Apollo 12 *Intrepid*	USA 2
3	Alan Bean	Apollo 12 *Intrepid*	USA 3
4	Alan Shepard	Apollo 14 *Antares*	USA 4

Pad 39A, Kennedy Space Centre, 28 November 1983. Spacelab 1. The ninth flight of the space shuttle – which came on the scene in April 1981 after many delays, ending the dominance of space by the Soviet Union – carried the European laboratory Spacelab. On board, four scientists worked for ten days carrying out over 70 experiments. Also on board were a pilot and the commander, John Young, making his sixth spaceflight. Young was commander of the first shuttle mission. (*NASA*)

Inside Shuttle *Challenger*, October 1984. Five men, two ladies. A record seven-person crew in space, including an oceanographer, a geologist, a physicist and a Canadian, demonstrating how accessible space has become to a wide range of people with different disciplines and illustrating how today we are exploiting space, not exploring it. (*NASA*)

Earth orbit, November 1984. Capture. During the 14th shuttle mission, satellites were captured by spacewalking astronauts and returned to Earth in a perfect demonstration of the value of man in space. (*NASA*)

EVA	Name	Craft	Country Placing
5	Edgar Mitchell	Apollo 14 *Antares*	USA 5
6	David Scott	Apollo 15 *Falcon*	USA 6
7	James Irwin	Apollo 15 *Falcon*	USA 7
8	John Young	Apollo 16 *Orion*	USA 8
9	Charles Duke	Apollo 16 *Orion*	USA 9
10	Eugene Cernan	Apollo 17 *Challenger*	USA 10
11	Jack Schmitt	Apollo 17 *Challenger*	USA 11
12	Owen Garriott	Skylab 3	USA 12
13	Jack Lousma	Skylab 3	USA 13
14	William Pogue	Skylab 4	USA 14
15	Gerald Carr	Skylab 4	USA 15
16	Edward Gibson	Skylab 4	USA 16
17	Vladimir Lyakhov	Soyuz T-9 Salyut 7	USSR 1
18	Aleksander Aleksandrov	Soyuz T-9 Salyut 7	USSR 2
19	Bruce McCandless	STS 41B *Challenger* MMU	USA 17
20	Robert Stewart	STS 41B *Challenger* MMU	USA 18
21	George Nelson	STS 41C *Challenger*	USA 19
22	James van Hoften	STS 41C *Challenger* MMU	USA 20
23	Leonid Kizim	Soyuz T-10 Salyut 7	USSR 3
24	Vladimir Solovyov	Soyuz T-10 Salyut 7	USSR 4
25	Joe Allen	STS 51A *Discovery*	USA 21
26	Dale Gardner	STS 51A *Discovery* MMU	USA 22

Musgrave's EVA. (*NASA*)

EVA log: three times

EVA	Name	Craft	Country Placing
1	David Scott	Apollo 15 *Falcon*	USA 1
2	James Irwin	Apollo 15 *Falcon*	USA 2
3	John Young	Apollo 16 *Orion*	USA 3
4	Charlie Duke	Apollo 16 *Orion*	USA 4
5	Eugene Cernan	Apollo 17 *Challenger*	USA 5
6	Jack Schmitt	Apollo 17 *Challenger*	USA 6
7	Charles Conrad	Skylab 2	USA 7
8	Owen Garriott	Skylab 3	USA 8
9	Gerald Carr	Skylab 4	USA 9
10	Edward Gibson	Skylab 4	USA 10
11	Vladimir Lyakhov	Soyuz T-9 Salyut 7	USSR 1
12	Leonid Kizim	Soyuz T-10 Salyut 7	USSR 2
13	Vladimir Solovyov	Soyuz T-10 Salyut 7	USSR 3

EVA log: four times

EVA	Name	Craft	Country Placing
1	Eugene Cernan	Apollo 17 *Challenger*	USA 1
2	Charles Conrad	Skylab 2	USA 2
3	Leonid Kizim	Soyuz T-10 Salyut 7	USSR 1
4	Vladimir Solovyov	Soyuz T-10 Salyut 7	USSR 2

EVA log: five and six times

EVA	Name	Craft	Country Placing
1	Leonid Kizim	Soyuz T-10 Salyut 7	USSR 1
2	Vladimir Solovyov	Soyuz T-10 Salyut 7	USSR 2

MMU Log

Flight	Name	MMU No	Duration h	Duration min	Mission EVA
STS 41B	McCandless	3	1	22	1
	Stewart	3	1	9	1
	McCandless	2		47	2
	Stewart	2		44	2
	McCandless	3	1	8	2
STS 41C	Nelson	2		42	1
	van Hoften	3		28	2
STS 51A	Allen	—		—	1
	Gardner (D)			—	2

Spacemen with over 5 hours' full EVA experience

Name	Country	Time h	min
Cernan	USA	24	12
Kizim	USSR	22	50*
Solovyov	USSR	22	50*
Schmitt	USA	22	5
Young	USA	20	14
Duke	USA	20	14
Scott	USA	17	36
Irwin	USA	17	36
Carr	USA	15	48
Gibson	USA	15	17
Garriott	USA	13	44
Pogue	USA	13	31
Conrad	USA	12	19
Allen	USA	11	42
Gardner	USA	11	42
McCandless	USA	11	37
Stewart	USA	11	37
Lousma	USA	10	59
Bean	USA	9	50
Shepard	USA	9	24
Mitchell	USA	9	24
Nelson	USA	9	13
Van Hoften	USA	9	13
Lyakhov	USSR	7	18
Aleksandrov	USSR	5	45

*Approximate times.

Stand-up EVAs (partial exposure)

Name	Craft
Michael Collins	Gemini 10
Richard Gordon	Gemini 11
Edwin Aldrin	Gemini 12 (2)
David Scott	Apollo 9 *Gumdrop*
David Scott	Apollo 15 *Falcon*
James Irwin	Apollo 15 *Endeavour*
Charles Duke	Apollo 16 *Casper*
Jack Schmitt	Apollo 17 *America*
Paul Weitz	Skylab 2

Bob Stewart in his MMU. (*NASA*)

The 'Astroflights' of the X-15 Rocket Plane

The Fédération Aéronautique Internationale (FAI) recognizes a space flight as one that exceeds an altitude of 100 km or 62 miles.

However, between 1962 and 1968 the American rocket plane, the X-15, flew 13 missions over a height of 50 miles. The military service pilots who flew some of these missions were awarded USAF astronaut wings for exceeding 50 miles, an altitude that had hitherto unofficially been designated as the beginning of 'space'.

A NASA pilot, Joe Walker, actually exceeded a height of 62 miles on two missions of the X-15 and so theoretically it can be claimed that he was an official spaceman.

None of these X-15 'astroflights' pilots has been officially recognized as a spaceman. However, no record of manned spaceflight is complete without a log of these flights.

The X-15 rocket plane was a hypersonic research vehicle flown 199 times between 1959 and 1968. There were three vehicles. Twelve men flew X-15s and two of them are official spacemen today – Neil Armstrong and Joe Engle (although only Engle flew an 'astroflight').

X-15 astroflight log

Date	Craft no.	Craft mission no.	Overall X-15 mission no.	Altitude (m)	Pilot	No. of flights
17 July 1962	X-15 3	7	62	59.16	Maj. Robert White, USAF	16
17 Jan. 1963	X-15 3	14	77	51.00	Joseph Walker, NASA	25
27 June 1963	X-15 3	20	87	55.00	Maj. Robert Rushworth, USAF	34
19 July 1963	X-15 3	21	90	65.30	Joseph Walker, NASA	25
22 Aug. 1963	X-15 3	22	91	66.75	Joseph Walker, NASA*	25
29 June 1965	X-15 3	44	138	53.14	Capt. Joseph Engle, USAF	16
10 Aug. 1965	X-15 3	46	143	51.70	Capt. Joseph Engle, USAF	16
28 Sep. 1965	X-15 3	49	150	56.00	John 'Jack' McKay, NASA	29
14 Oct. 1965	X-15 1	61	153	50.17	Capt. Joseph Engle, USAF	16
1 Nov. 1966	X-15 3	56	174	58.00	William Dana, NASA	16
17 Oct. 1967	X-15 3	64	190	53.40	Maj. William 'Pete' Knight, USAF	16
15 Nov. 1967	X-15 3	65	191	50.40	Maj. Michael Adams, USAF†	7
21 Aug. 1968	X-15 1	79	197	50.70	William Dana, NASA	16

*Joe Walker was killed in mid-air crash, 8 June 1966.
†Mike Adams was killed on his X-15 mission.

Other X-15 pilots*

Name	No. of flights	History
Ivan Kincheloe	—	Killed in air crash before flying
Scott Crossfield	14	Former X rocket plane ace
Forrest Petersen	5	Navy pilot
Neil Armstrong	7	Ex Navy pilot with NACA then NASA
Milton Thompson	14	Former Dyna Soar astronaut

*These did not make astroflights.

The X-15 rocket plane is launched from its B-52 mother-ship in November 1965. (*USAF*)

All Known Soviet Cosmonauts

Flight-experienced cosmonauts are detailed in A–Z

March 1960 (20 military pilots)

Flights

Col. Yuri Gagarin
Lt Gen. Gherman Titov
Maj. Gen. Andrian Nikolyev
Maj. Gen. Pavel Popovich
Col. Valeri Bykovsky
Col. Eng. Vladimir Komarov
Col. Pavel Belyayev
Maj. Gen. Alexei Leonov
Col. Boris Volynov
Col. Eng. Yevgeni Khrunov
Maj. Gen. Georgi Shonin
Maj. Gen. Viktor Gorbatko

No flights

Col. 'Dmitri'	Voskhod 2 back-up commander; retired, 1969
Col. 'Grigori'	Original Vostok pilot
Col. 'Anatoli'	Original Vostok 3 pilot; retired, 1962
Col. 'Ivan'	Retired, 1962
Col. 'Mars'	Retired, 1962
Col. 'Valentin' 1	Retired, 1960
Col. 'Valentin' 2	Retired, 1962
Col. 'Valentin' 3	Retired, 1962

Gagarin congratulates Leonov after the first spacewalk. (*Novosti*)

Gagarin. (*Novosti*)

February 1962 (associated with Vostok 6)

Flights

Col. Eng. Valentina Tereshkova

No flights

Tanya Tortchillova	Back-up Vostok 6 pilot

It has been suggested by some sources that Tortchillova was the original prime pilot but was indisposed.

January 1963 (military pilots/flight engineers)

Flight commanders	Flight engineers
Lt Gen. Vladimir Shatalov	Col. Eng. Yuri Artyuhkin
Maj. Gen. Anatoli Filipchenko	Col. Lev Demin
Lt Col. Georgi Dobrovolsky	Col. Eng. Vitali Zholobov
Maj. Gen. Alexei Gubarev	

1964 (associated with Voskhod)

Flights

Lt Boris Yegerov
Konstantin Feoktistov
Col. Vasili Lazarev
Oleg Makarov

Gagarin, Komarov and Nikolyev. (*Novosti*)

February 1964

Flights

Lt Gen. Georgi Beregovoi

1965 (military flight engineers)

Flights

Col. Valeri Rozhdestvensky
Col. Eng. Yuri Glazkov

October 1965 (military pilots)

Flights

Maj. Gen. Pyotr Klimuk
Col. Gennadi Serafanov
Col. Vyacheslav Zudov
Col. Leonid Kizim

August 1966 (civilian flight engineers)

Flights

Alexei Yeliseyev
Valeri Kubasov
Vladislav Volkov

1966 (military flight engineers)

No flights

Col. Sergei Anokhin	Retired due to ill health

January 1967 (civilian flight engineers)

Flights

Vitali Sevastyanov
Nikolai Ruchavishnikov
Georgi Grechko

1967 (military pilots)

Flights

Col. Vladimir Kovalyonok
Col. Vladimir Lyakhov
Col. Yuri Malyshev

1969 (civilian flight engineers)

Flights

Viktor Patsayev

1970 (civilian flight engineers)

Flights

Aleksander Ivanchenkov

No flights

Boris Andreyev	A Soyuz 19 back-up flight engineer

1970 (military pilots)

Flights

Col. Vladimir Dzhanibekov
Col. Yuri Romanenko
Col. Anatoli Berezovoi
Col. Leonid Popov

No flights

Lt Col. Valeri Illarionov	ASTP back-up

1972 (civilian flight engineers)

Flight

Valentin Lebedev

Voskhod 3 back-up crew of Shatalov (foreground) and Beregovoi. (*Novosti*)

Vostok 1 to Soyuz 8 cosmonauts (except Feoktistov). (*Novosti*)

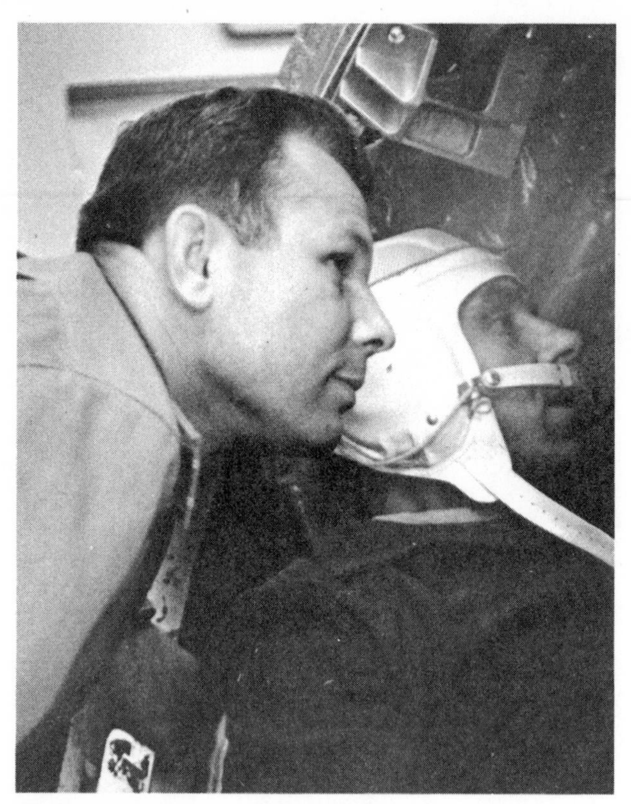

Gagarin with Komarov, the back-up and prime crew of Soyuz 1. (*Novosti*)

1973 (civilian flight engineers)

Flights

Vladimir Aksyonov
Valeri Ryumin
Gennadi Strekalov

1976 (military pilots)

Flight

Col. Vladimir Titov

1977 (doctors)

Flight

Dr Oleg Atkov

Soyuz T-3 crew, left to right, Kizim, Makarov and Strekalov. (*Novosti*)

1978 (civilian flight engineers)

Flights

Viktor Savinykh
Vladimir Solovyov
Alexander Aleksandrov
Alexander Serebrov
Igor Volk

1980 (female crew)

Flights

Svetlana Savitskaya

No flights

Three others selected

1982 (female crew)

No flights

Six more selected?

The original Soyuz 2 crew who were to dock with Soyuz 1 and two of whom were to transfer to it by spacewalking. They are, left to right, Yevegni Khrunov, Valeri Bykovsky and Alexei Yelesyev. (*Novosti*)

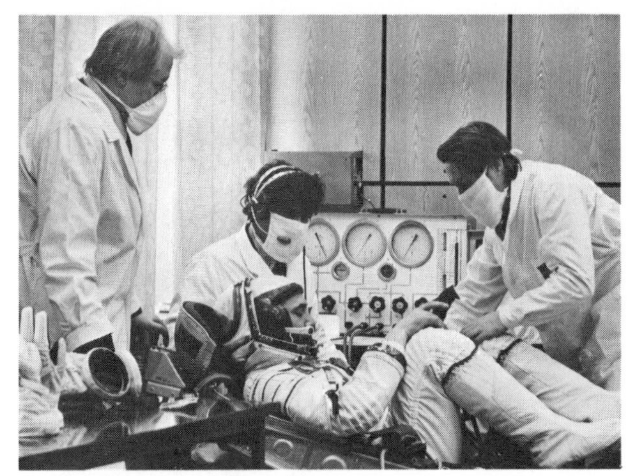

Training for Soyuz. (*Novosti*)

Intercosmos Soyuz cosmonaut researchers

Flight (see A–Z)

Vladimir Remek	Czechoslovakia
Miroslaw Hermaszewski	Poland
Sigmund Jähn	East Germany
Georgi Ivanov	Bulgaria
Bertalan Farkas	Hungary
Pham Tuan	Vietnam
Arnaldo Mendez	Cuba
Jugderdemidyin Gurragcha	Mongolia
Dumitru Prunariu	Romania
Jean Loup Chrétien	France
Rakesh Sharma	India

Back-ups – no flights

Oldrzich Pelczak	Czechoslovakia
Zenon Jankowski	Poland
Eberhard Kollner	East Germany
Alexander Alexandrov	Bulgaria
Bela Magyari	Hungary
Bui Thanh Liem	Vietnam*
Jose Armando Lopez Falcon	Cuba
Maydarjaviyn Gonzorig	Mongolia
Dumitru Dediu	Romania
Patrick Baudry	France†
Ravish Malhotra	India

*Killed in air crash, 1981.
†Space Shuttle payload specialist.

Soyuz T-11 back up crew: Malhotra (India), Berezovoi and Grechko. (*Novosti*)

NASA Astronauts

Flight experienced astronauts see A–Z

Mercury 7 – April 1959 (7)

Flights

Scott Carpenter	
Gordon Cooper	
John Glenn	
Walter Schirra	
Alan Shepard	
Deke Slayton	No Mercury flight; ASTP
Gus Grissom	Died, 27 January 1967 in Apollo 1 fire

The Mercury team in 1959. Left to right, Grissom, Shepard, Carpenter, Schirra, Slayton, Glenn and Cooper. (*NASA*)

Group 2 – September 1962 (9)

Flights

Neil Armstrong	
Frank Borman	
Charles Conrad	
James McDivitt	
James Lovell	
Thomas Stafford	
Edward White	Died, 27 January 1967, in Apollo 1 fire
John Young	**Active**

No flights

Elliott See	Gemini 5 back-up pilot; Gemini 9 commander; killed in air crash, 28 February 1966

Group 2 astronauts, left to right, White, Stafford, Lovell, Borman and See. (*NASA*)

Group 1 and 2 astronauts on emergency survival course. (*NASA*)

Group 3 – October 1963 (14)

Flights

William Anders
Edwin Aldrin
Alan Bean
Eugene Cernan
Michael Collins
Walter Cunningham
Donn Eisele
Richard Gordon
Russell Schweickart
David Scott

No flights

Maj. Charles Bassett, USAF	Gemini 9 pilot; killed in air crash, 28 February 1966
Lt Cdr. Roger Chaffee, USN	Apollo 1 pilot; killed, 27 January 1967
Capt. Theodore Freeman, USAF	Killed in air crash, 31 October 1964
Capt. Clifton Williams, USMC	Gemini 10 back-up pilot; Apollo 3 back-up lunar module pilot; killed in air crash, 5 October 1967

The Group 6 astro scientists in 1967. Some gave up hope and resigned but others persevered and made it into space. Left to right: Allen, Henize, Chapman, England, Parker, Holmquest, Thornton, Musgrave, Llewellyn, Lenoir and O'Leary. (*NASA*)

Group 4 – June 1965 (6 astro-scientists)

Flights

Edward Gibson	
Owen Garriott	**Active**
Joseph Kerwin	**Active**
Harrison Schmitt	

No flights

Curtis Michel	Doctorate in astrophysics, resigned, 1969
Duane Graveline, MD	Resigned, 1965

Group 5 – April 1966 (19)

Flights

Vance Brand	**Active**
Gerald Carr	
Charles Duke	
Joe Engle	**Active**
Ronald Evans	
Fred Haise	
James Irwin	
Don Lind	**Active**
Jack Lousma	
Ken Mattingly	
Bruce McCandless	**Active**
Edgar Mitchell	
William Pogue	
Stuart Roosa	
Jack Swigert	Died 28 December 1982
Paul Weitz	**Active**
Alfred Worden	

No flights

Lt Cdr John Bull, USN Rtd	Resigned, 1968 due to ill health
Maj. Edward Givens, USAF	Killed in car crash, 6 June 1967

Group 6 – August 1967 (11 astro-scientists)

Flights

Joe Allen	**Active**
William Lenoir	
Story Musgrave	**Active**
Robert Parker	**Active**
William Thornton	**Active**

No flights

Philip Chapman	Doctorate in aeronautics and astronautics; resigned, 1972
Anthony England, MD	**Active**; STS 51F mission specialist
Karl Henize, astronomer	**Active**; STS 51F mission specialist
Donald Holmquest, MD	Resigned, 1973
John Llewellyn	Doctorate in chemistry; resigned, 1967*
Brian O'Leary, astronomer	Resigned, 1968

*Born in Cardiff; US citizen, 1966

Group 7 August 1969 (7 former MOL astronauts)

Flights

Karol Bobko	**Active**
Robert Crippen	**Active**
Gordon Fullerton	**Active**
Henry Hartsfield	**Active**
Robert Overmyer	**Active**
Donald Peterson	
Richard Truly	

Group 8 – August 1978 (35 Shuttle pilots and mission specialists)

Pilots
Flights

Daniel Brandenstein	**Active**
Michael Coats	**Active**
Robert Gibson	**Active**†
Fred Gregory	**Active**
David Griggs	**Active**
Rich Hauck	**Active**
Jon McBride	**Active**
Dick Scobee	**Active**
Brewster Shaw	**Active**
Loren Shriver	**Active**
David Walker	**Active**
Donald Williams	**Active**

No flights

Lt Col. Richard Covey, USAF	**Active** STS 51I pilot
Cdr John Creighton, USN	**Active** STS 51G
Lt Col. Steven Nagel, USAF	**Active**; STS 51G mission specialist STS 61A pilot

Mission specialists
Flights

Guion Bluford	**Active**
James Buchli	**Active**
John Fabian	**Active**
Anna Fisher	**Active**†
Dale Gardner	**Active**
Terry Hart	
Steven Hawley	**Active**†
Jeff Hoffman	**Active**
Ronald McNair	**Active**
Richard Mullane	**Active**
George Nelson	**Active**
Ellison Onizuka	**Active**
Judith Resnik	**Active**
Sally Ride	**Active**†
Rhea Seddon	**Active**†
Robert Stewart	**Active**
Kathyrn Sullivan	**Active**
Norman Thagard	**Active**
James van Hoften	**Active**

No flight

Shannon Lucid Dr Biochemistry	**Active**; STS 51G

Group 9 – May 1980 (19 Shuttle pilots and mission specialists)

Pilots
No flights

Col. John Blaha, USAF	**Active**; STS 61H pilot
Lt Col. Charles Bolden, USMC	**Active**; STS 61C pilot
Col. Roy Bridges, USAF	**Active**; STS 51F pilot
Lt Col. Guy Gardner, USAF	**Active**; STS 62A pilot
Lt Col. Ronald Grabe, USAF	**Active**; STS 51J pilot
Maj. Bryan O'Connor, USMC	**Active**; STS 61B pilot
Cdr Richard Richards, USN	**Active**; STS 61E pilot
Cdr Mike Smith, USN	**Active**; STS 51L pilot

Mission specialists
Flights

David Leestma MS	**Active**

No flights

James Bagian, MD	**Active**; STS 61D
Franklin Chang, Dr of physics	**Active**; STS 61C
Mary Cleave, Dr of civil engineering	**Active**; STS 61B
Bonnie Dunbar	**Active**; STS 61A
William Fisher, MD	**Active**†; STS 51I
Maj. David Hilmers, USMC	**Active**; STS 51J
John Lounge, astrophysicist	**Active**; STS 51G
Maj. Jerry Ross, USAF	**Active**; STS 61B, STS 62A
Lt Col. Sherwood Spring, US Army	**Active**; STS 61B
Lt Col. Robert Springer, USMC	**Active**; STS 61D

(Claude Nicollier and Wubbo Ockels, ESA astronauts, were added to this group and became NASA mission specialists.)

Group 10 – May 1984 (17 Shuttle pilots and mission specialists)

Pilots

No flights

Capt. Mark Brown, USAF	**Active**
Cdr Manley Carter, USN	**Active**
Ltd Cdr Frank Culbertson, USN	**Active**
Capt. Sidney Gutierrez, USN	**Active**
Capt. Lloyd Hammond, USAF	**Active**
Capt. Mark Lee, USAF	**Active**
Lt James Wetherbee, USN	**Active**

Mission specialists

No flights

Maj. James Adamson, US	**Active**
Maj. Kenneth Cameron, USMC	**Active**
Lt Col. John Casper, USAF	**Active**
Marsha Ivins	**Active**
George Low	**Active**
Lt Cdr Michael McCulley, USN	**Active**
Lt Cdr William Shepherd, USN	**Active**
Ellen Shulman, MD	**Active**
Kathryn Thornton	**Active**
Charles Veach	**Active**

(Group 10 are essentially candidates and not on flight status until June 1985.)

Total selected 144*

Total active 72* (plus Group 10/17 = 89)

†Husband and wife teams
Robert Gibson Rhea Seddon
Steven Hawley Sally Ride
William and Anna Fisher

*ESA astronauts Claude Nicollier and Wubbo Ockels qualified as NASA mission specialists and are theoretically available to fly non-ESA flights. Nicollier was named as MS on EOM1/STS 51H, a reflight of some Spacelab 1 experiments. This flight has been cancelled but is planned later as STS 61K, but Nicollier has not yet been named.

The Group 10 astronauts. (*NASA*)

The Dyna Soar astronauts, clockwise, Wood, Thompson, Crews, Gordon, Rogers and Knight. (*USAF*)

X-20 Dyna Soar astronauts September 1962

In 1962 plans existed for a Space Shuttle glider called Dyna Soar to be launched by a Titan 3 into orbit. These six pilots were selected to the programme which was cancelled in 1963.

Capt. Albert Crews, USAF	
Capt. William 'Pete' Knight, USAF†	
Milton Thompson†	
Capt. Henry Gordon, USAF	
Capt. Russell Rogers, USAF	Deceased*
Maj. James Wood, USAF	

*Russell Rogers was killed on 13 September 1967 when his F-105 exploded in mid-air near Okinawa.
†Pete Knight went on to fly the X-15 into 'space' and Milton Thompson also flew the rocket plane (see X-15 tables).

Manned orbital laboratory astronauts

This military project involved the use of a Gemini spacecraft attached to a cylindrical module on top of a Titan 3C booster. Missions lasting 30 days were envisaged. The project was cancelled in 1969.

Group 1 – 12 November 1965

Maj. Michael Adams, USAF	Went on to fly X-15
Maj. Albert Crews, USAF	
Ltd Cdr John Finley, USN	
Capt. Richard Lawyer, USN	
Maj. Lachlan MacLeay, USAF	
Capt. Francis Greg Neubeck, USAF	
Maj. James Taylor, USAF	Killed in air crash, 4 September 1970
Lt Richard Truly, USN	Joined NASA

Group 2 – 17 June 1966

Capt. Karol Bobko, USAF	Joined NASA

Lt Robert Crippen, USN	Joined NASA
Capt. C. Gordon Fullerton, USAG	Joined NASA
Capt. Henry Hartsfield, USAF	Joined NASA
Capt. Robert Overmyer	Joined NASA

Group 3 – 30 June 1967

Maj. James Abrahamson	Now chief 'Star Wars'; former Shuttle chief
Lt Col. Robert Herres, USAF	
Maj. Robert Lawrence, USAF	The first negro astronaut; killed in air crash, 8 December 1967
Maj. Donald Peterson, USAF	Joined NASA

The MOL 1 and 2 astronauts. Top row, left to right, Taylor Neubeck, Crews, MacLeay, Finley, Truly. Bottom row, left to right, Crippen, Overmyer, Bobko, Fullerton and Hartsfield. (*USAF*)

Space Shuttle Passenger List

(With date of selection)

A fourth group of crew persons on shuttle flights are non-NASA astronaut personnel called payload specialists who fly on board, mainly to operate scientific equipment or to help deploy satellites. Others fly as 'passenger observers' such as a government official and a US teacher. Later, journalists photographers and artists, among others may fly.

Spacelab 1 STS 9

Candidates from European Space Agency, 1977

Franco Malerba	Italy
Ulf Merbold	West Germany
Claude Nicollier	Switzerland
Wubbo Ockels	Netherlands

Malerba was eliminated in 1978 and Nicollier and Ockels qualified as NASA mission specialists available for non-ESA flights as well, if required. Merbold did not, but he was selected as prime payload specialist on Spacelab 1 with Ockels as his back-up.

Candidates from USA, 1977

Craig Fisher
Michael Lampton
Byron Lichtenberg
Robert Menzies
Ann Whitaker
Richard Terrile

Lampton and Lichtenberg were retained in 1978 and Lichtenberg was chosen as prime payload specialist for Spacelab (with Lampton back-up).

Spacelab 2 STS 51F
US payload specialists, 1978

Loren Acton
John David Bartoe
Dianne Prinz
George Simon

Acton and Bartoe were chosen as prime crew with Prinz and Simon as back-ups.

Spacelab 3 STS 51B
US payload specialists, 1983

Eugene Trinh
Taylor Wang
Mary Helen Johnston
Lodewijk van den Berg

Wang and Van den Berg were chosen as prime crew with Trinh and Johnston as back-ups.

Spacelab 4 SLS-1 STS 61D
US payload specialists, 1983

Millie Hughes-Fulford	
Francis Gaffney	
Robert Phillips	
Bill Williams	Resigned, 1985

Gaffney and Phillips selected. Hughes-Fulford named for SLS-2, STS 71G

Canadian payload specialists, left to right: Money, Thirsk, Bondar, MacLean, Tryggvason and Garneau. (*Research Council of Canada*)

Spacelab D-1 STS 61A European payload specialists, 1982

Wubbo Ockels	
Ulf Merbold	
Claude Nicollier*	
Reinhard Furrer	West Germany
Ernst Willi Messerschmidt	West Germany

Furrer, Messerschmidt and Ockels are prime crew with Merbold back-up to Ockels.

*Despite being NASA MS

McDonnell Douglas payload specialists, 1983-1985

Charles Walker	STS 41D, STS 51D and 61B

Test engineer of space processing machine.

New McDonnell Douglas PS Robert Wood will fly in 1986.

Canadian payload specialist astronauts, 1984

Marc Garneau	STS 41G
Roberta Bondar	
Bjarni Tryggvason	
Steve MacLean	
Bob Thirsk	STS 41G back-up
Ken Money	

Spacelab D-1 payload specialists, left to right, Merbold, Ockels (German official), Furrer and Messerschmidt. (*DFVLR*)

Oceanographer payload specialist, 1984

Paul Scully-Power	STS 41G

Hughes Communications Inc payload specialists, 1984

Gregory Jarvis	STS 51I
John Konrad	No Flight yet
William Butterworth	Alternate
Steven Cunningham	Alternate

Astro payload specialists, 1984

(For three astronomy shuttle missions – first mission STS 61E in March 1986; second mission possible; third now unlikely.)

Samual Durrance
Kenneth Nordsieck
Ronald Parise

Two to fly as Astro 1, the remaining PS on Astro 2.

Skynet 4 UK payload specialists, 1984

Cdr Peter Longhurst	(Navy)	STS 71D Prime
Lt. Col. Anthony Boyle	(Army)	
Sq. Ld. Nigel Wood	(RAF)	STS 61H Prime
Christopher Holmes	(MOD)	STS 71D Back-up
Boyle was dropped and replaced by Lt. Col. Richard Farrimond	(Army)	STS 61H Back-up

French payload specialists, 1984

Patrick Baudry	STS 51G
Jean Loup Chrétien	Back-up 51G*

*Cosmonaut researcher on Soyuz T-6

Earth Observation Mission payload specialists 1984

Selected for EOM1/61K.

Byron Lichtenberg
Michael Lampton

(Note: Claude Nicollier was named mission specialist on this mission.)

DoD manned spaceflight engineers 1982

Twenty-five military officers were selected for DoD shuttle flights. The first, Gary Payton, was assigned to STS 10 in 1983. The flight eventually became STS 51C in 1985. Flights 51J and 62A are also scheduled to carry manned space flight engineers.

Other payload specialists

Senator Edwin Jake Garn, Rep Utah, STS 51D

A US teacher and back-up to be selected late 1985 for 1986 flight, STS 51L.

Congressman Bill Nelson of Florida is a probable candidate for a flight in 1986.

An American journalist is expected to fly in 1986/87.

RCA Astro Electronics is appointing a payload specialist for mission 61C.

American Satellite Communications Co. is appointing a payload specialist for 61L.

National payload specialists

Japan
Indonesia
China
Saudi Arabia
India
Mexico

Each of the above countries is flying its own individual payload specialist on shuttle missions.
Name of Saudi Arabian for mission 51G is Sultan Salman Abdelazize al-Saud, with back up Abdulmohsen Hamad Al-Bassam.
Mexico is selecting PS for 61B.
Indian flies on 61I.
Indonesian flies in 61H.

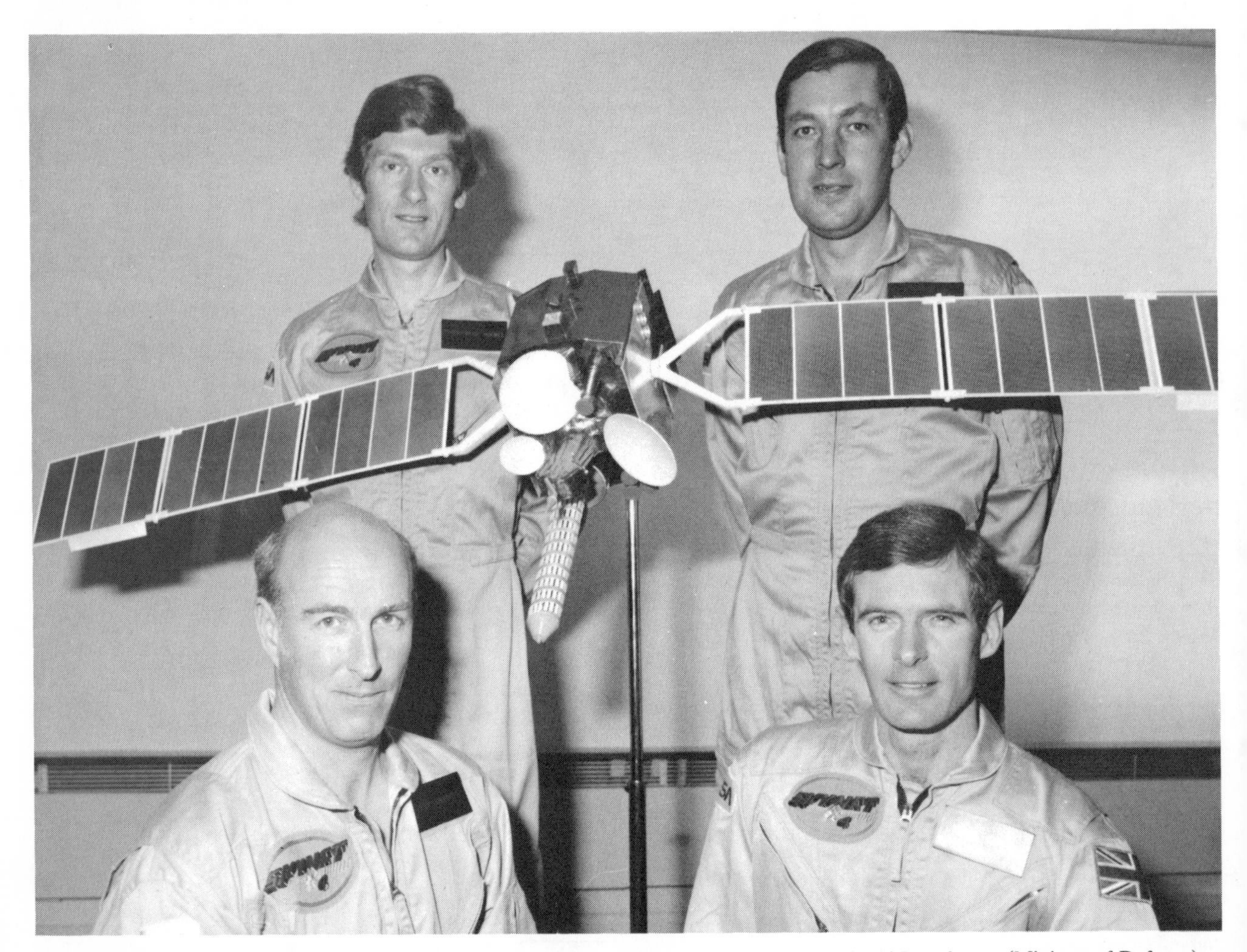

The British Skynet Payload Specialists, clockwise, from top left, Holmes, Farrimond, Wood and Longhurst, (Ministry of Defence).

NASA Flight Crew Selections

(Note: the year does not necessarily denote the year that selection was made **official**.)
Actual flights set in **bold.**

Year	Flight	Crew	Notes
1961	MR3	Shepard Grissom Glenn	First candidates. Original intention was to fly all 7 astronauts on Redstone flights
	MR3	**Shepard**	Prime sub-orbital
		Glenn	Back-up
	MR4	**Grissom**	Sub-orbital
		Glenn	
	MR5	Glenn	Final Redstone flight cancelled
	MA6	**Glenn**	Orbital
		Carpenter	
	MA7	Slayton	Orbital. Original choice for first orbital flight had MR5 been retained
		Schirra	
1962		Slayton	Dropped from status with heart flutter
	MA7	**Carpenter**	Orbital
		Schirra	
	MA8	**Schirra**	Orbital
		Cooper	
	MA9	**Cooper**	Orbital
		Shepard	
1963	MA10	Shepard	Planned 3-day orbital flight cancelled
	Gemini 3	Shepard	
		Borman or	
		Stafford	
		Shepard	Grounded by Menière's Disease
1964	**Gemini 3**	**Grissom**	Command pilot orbital
		Young	Pilot
		Schirra	
		Stafford	
	Gemini 4	**McDivitt**	Spacewalk
		White	
		Borman	
		Lovell	
1965	**Gemini 5**	**Cooper**	Long duration
		Conrad	
		Armstrong	
		See	
	Gemini 6	**Schirra**	Rendezvous and docking with Agena
		Stafford	Became rendezvous with Gemini 7
		Grissom	
		Young	
	Gemini 7	**Borman**	(Launched before Gemini 6) Long duration
		Lovell	
		White	
		Collins	
	Gemini 8	**Armstrong**	Rendezvous and docking and spacewalk, not completed because of emergency landing
		Scott	
		Conrad	
		Gordon	

	Gemini 9	See	Rendezvous, docking spacewalk
		Bassett	
		Stafford	
		Cernan	
1966	Gemini 10	Young	Rendezvous, docking, spacewalk
		Collins	
		Lovell	
		Aldrin	
			See and Bassett killed in air crash, 28 February 1966
	Gemini 9	**Stafford**	Rendezvous, docking, spacewalk
		Cernan	
		Lovell	
		Aldrin	
	Gemini 10	**Young**	Rendezvous, docking, spacewalk
		Collins	
		Bean	
		Williams	
	Gemini 11	**Conrad**	Rendezvous, docking, spacewalk
		Gordon	
		Armstrong	
		Anders	
	Gemini 12	**Lovell**	Rendezvous, docking, spacewalk
		Aldrin	
		Cooper	
		Cernan	
	Apollo 1	Grissom	Orbital test of CM
		White	
		Chaffee	(Originally Eisele but he broke his arm)
		McDivitt	
		Scott	
		Schweickart	
	Apollo 2 (A)	Schirra	Repeat of Apollo 1
		Eisele	Apollo 2 (A) was cancelled
		Cunningham	
		Borman	
		Stafford	
		Collins	
	Apollo 1	Grissom	Cdr Orbital test of CM
		White	Senior pilot
		Chaffee	Pilot
		Schirra	
		Eisele	
		Cunningham	
	Apollo 2 (B)	McDivitt	Rendezvous, docking with LM launched separately by Saturn 1B
		Scott	
		Schweickart	
		Stafford	
		Young	
		Cernan	
	Apollo 3	Borman	High altitude LM test. Launched by Saturn 5
		Collins	
		Anders	
		Conrad	
		Gordon	
		Williams	

Young at the time of his selection for Gemini 3. (*NASA*)

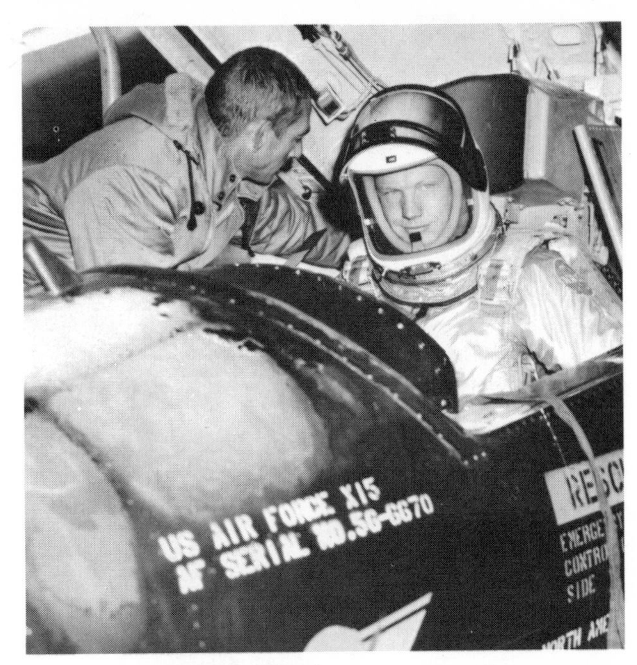

X-15 pilot Armstrong. (*NASA*)

Year	Mission	Crew	Notes
1967	Apollo 1		Spacecraft fire, 27 January 1967
	Apollo 7	**Schirra**	Cdr CM in Earth orbit
	AS-205	**Eisele**	Senior pilot
		Cunningham	Pilot
		Stafford	
		Young	
		Cernan	
	AS-504	McDivitt	LM test in Earth orbit
		Scott	
		Schweickart	
		Conrad	
		Gordon	
		Bean	(Williams killed in air crash)
	AS-505	Borman	LM test in deep space
		Collins	
		Anders	
		Armstrong	
		Lovell	
		Aldrin	
1968			Michael Collins had surgery on his back and Apollo LM was delayed
	Apollo 8	**Borman**	Commander – CM in lunar orbit
		Lovell	Senior pilot
		Anders	Pilot
		Armstrong	
		Aldrin	
		Haise	
	Apollo 9	**McDivitt**	Commander – LM in Earth orbit
		Scott	Command module pilot
		Schweickart	LM pilot
		Conrad	
		Gordon	
		Bean	

Anders. *Textron*

Cunningham. *Capital Group*

	Apollo 10	**Stafford**	LM in lunar orbit
		Young	
		Cernan	
		Cooper	
		Eisele	
		Mitchell	
			Collins back on flight status
1969	**Apollo 11**	**Armstrong**	Lunar landing
		Collins	
		Aldrin	
		Lovell	
		Anders	
		Haise	
	Apollo 12	**Conrad**	Lunar landing
		Gordon	
		Bean	
		Scott	
		Worden	
		Irwin	
			Shepard restored to flight status; Cooper resigned
	Apollo 13	Lovell	Lunar landing
		Mattingly	(Anders resigned)
		Haise	
		Young	
		Swigert	
		Duke	
	Apollo 14	**Shepard**	Lunar landing
		Roosa	
		Mitchell	
		Cernan	(Would have been Collins had he not retired)
		Evans	
		Engle	

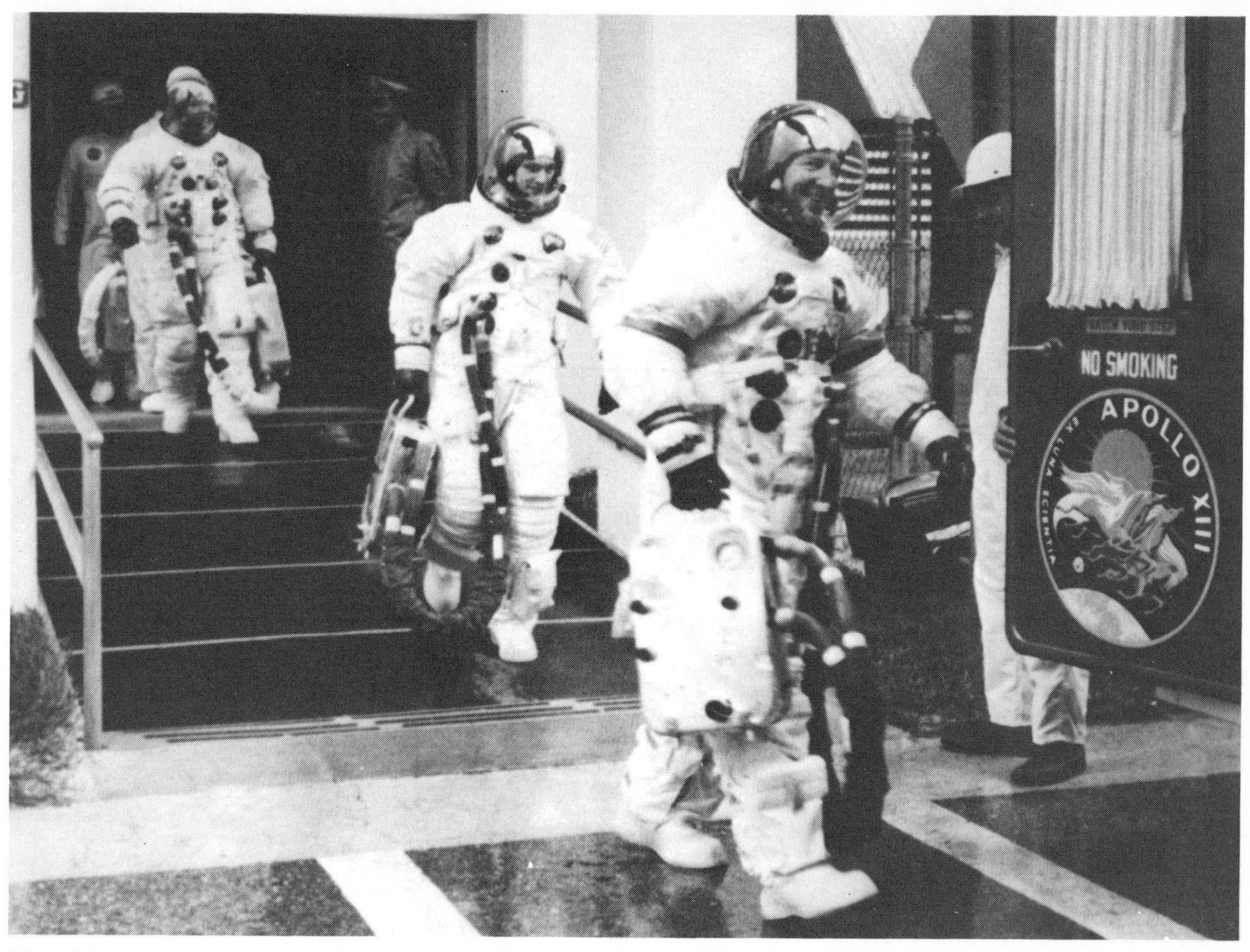

The original Apollo 13 crew, Lovell, right, Mattingly and Haise during rehearsal. (*NASA*)

Year	Mission	Crew	Notes
1970			Mattingly is German measles suspect 2 days before launch
	Apollo 13	**Lovell**	Lunar landing abort
		Swigert	
		Haise	
	Apollo 15	**Scott**	Lunar roving vehicle
		Worden	
		Irwin	
		Gordon	
		Brand	
		Schmitt	
	Apollo 16	Young	Lunar landing
		Mattingly	
		Duke	
		Haise	
		Pogue	
		Carr	
	Apollo 17	Cernan	Lunar landing
		Evans	
		Engle	
		Conrad	
		Weitz	
		Lousma	

	Apollo 18	Gordon	
		Brand	
		Schmitt	
	Apollo 19	Haise	Lunar landing
		Pogue	
		Carr	
	Apollo 20	Conrad	Lunar landing
		Weitz	
		Lousma	
	Skylab	Cunningham	Space station
		Schweickart	(Probable selection)
		Stafford	
		Kerwin	
		Garriott	
		Gibson	
		McCandless	(Probable selection)
		Lind	
		Eisele	
1971			Apollos 18, 19, 20 cancelled
	Apollo 16	**Young**	Lunar landing
		Mattingly	
		Duke	
		Haise	
		Roosa	
		Mitchell	
	Apollo 17	**Cernan**	Last lunar landing
		Evans	
		Schmitt	
		Scott	
		Worden	
		Irwin	
			Scott, Worden, Irwin dropped for disciplinary reasons replaced by:
		Young	
		Roosa	(Originally Mattingly)
		Duke	
	Skylab 2	**Conrad**	Cdr – Space station
		Kerwin	Science pilot
		Weitz	Pilot
		Schweickart	
		Musgrave	
		McCandless	
	Skylab 3	**Bean**	Space station
		Garriott	
		Lousma	
		Brand	
		Lenoir	
		Lind	
	Skylab 4	**Carr**	Space station
		Gibson	
		Pogue	
		Brand	
		Lenoir	
		Lind	
	Skylab Rescue	Brand	
		Lind	
		Schweickart	
		McCandless	

Year	Mission	Crew	Notes
1973	ASTP	Stafford	Joint Russian flight
		Swigert	
		Slayton	
			Swigert dropped
	ASTP	**Stafford**	Cdr
		Brand	Command module pilot
		Slayton	Docking module pilot
		Bean	
		Evans	
		Lousma	
1976	**ALT**	**Haise**	Shuttle approach and landing tests
		Fullerton	
		Engle	
		Truly	
1978	OFT 1	Young	Orbital flight tests of shuttle
		Crippen	
	OFT 2	Engle	
		Truly	
	OFT 3	Haise	Was to reboost Skylab into higher orbit
		Lousma	
	OFT 4	Brand	
		Fullerton	
1979			Haise resigns and Skylab re-enters
	OFT 3	Lousma	
		Fullerton	
	OFT 4	Brand	
		Overmyer	
			OFT becomes STS, schedule change
1980	**STS 1**	**Young**	Cdr – 1st test flight
		Crippen	Pilot
		Engle	
		Truly	
	STS 2	**Engle**	
		Truly	
		Mattingly	
		Hartsfield	
1981	**STS 3**	**Lousma**	
		Fullerton	
		Mattingly	
		Hartsfield	
	STS 4	**Mattingly**	
		Hartsfield	NASA drops back-ups
1982	**STS 5**	**Brand**	
		Overmyer	
		Allen	MS
		Lenoir	MS
	STS 6	**Weitz**	
		Bobko	
		Peterson	MS
		Musgrave	MS
	STS 7	**Crippen**	
		Hauck	
		Fabian	MS
		Ride	MS
		Thagard	MS (added later)

STS 8	**Truly**	
	Brandenstein	
	Bluford	MS
	Gardner	MS
	Thornton	MS (added later)
STS 9	**Young**	
	Shaw	
	Garriott	MS
	Parker	MS
	Merbold	Payload Specialist (PS)
	Lichtenberg	PS

ielson, wearing a Manned Manoeuvring Unit heads out towards Solar Max during the STS 41C repair mission. (*NASA*)

Shuttle re-schedules result in many designation changes

Year	Mission	Crew	Notes
1983	**STS** 10/41E/41H/**51C/(15)**	**Mattingly**	
		Shriver	
		Onizuka	MS
		Buchli	MS
		Payton	PS (added later)
	STS 41H	Bobko	Military standby crew
		Grabe	
		Mullane	
		Hilmers	
	STS (11) **41B (10)**	**Brand**	(New designation system begins)
		Gibson	
		McCandless	MS
		Stewart	MS
		McNair	MS
	STS (12) **41D (12)**	**Hartsfield**	
		Coats	
		Resnik	MS
		Mullane	MS
		Hawley	MS
		Walker, C.	PS (added later)
	STS (13) **41C (11)**	**Crippen**	
		Scobee	
		Hart	MS
		Nelson	MS
		van Hoften	MS
	STS (14) 41E/41F/41G/51E	Bobko	
		Williams	
		Seddon	MS
		Hoffman	MS
		Griggs	MS
		Baudry	PS (added later)
		Garn	PS (added later)
			STS 41D abort cancels 41F – crew reassigned to 51E. 51E mission scrubbed in March 1985. Crew reassigned to 51D except Baudry. Causes complete reshuffle of later assigned astronaut crews.
	STS (15) 41G/41E/**51A (14)**	**Hauck**	
		Walker, D.	
		Fisher, A.	MS
		Allen	MS
		Gardner	MS
	STS (16) **41G/(13)**	**Crippen**	
		McBride	MS
		Leestma	MS
		Ride	MS
		Sullivan	MS
		Garneau	PS (added later)
		Scully-Power	PS (added later)
1984	STS (17) 51A/51D	Brandenstein	Reassigned
		Creighton	Reassigned
		Nagel	MS reassigned
		Fabian	MS reassigned
		Lucid	MS reassigned
		Walker, C.	PS (added later) Retained for new 51D
		Jarvis	PS (added later) Reassigned

	STS (18) **51B Spacelab 3 (17)**	**Overmyer**	
		Gregory	
		Thagard	MS
		Thornton	MS
		Lind	MS
		van den Berg	PS
		Wang	PS
	STS (19) 51C/51G	Engle	Reassigned
		Covey	Reassigned
		Fisher, W.	MS Reassigned
		Lounge	MS Reassigned
		Buchli	MS replaced by van Hoften because of delay to his first flight. Reassigned
		Baudry	PS (moved to 51E, then back to new 51G crew)
		Saudi Arabian	PS Reassigned
	STS 51D/51I	Shaw	(Crew reassigned in total)
		O'Connor	
		Cleave	MS
		Spring	MS
		Ross	MS
	STS 51F Spacelab 2	Fullerton	
		Griggs	Replaced by Bridges due to proximity of what became 51E
		Musgrave	MS
		England	MS
		Henize	MS
		Acton	PS
		Bartoe	PS
	STS 51H/EOM1	Brand	(Flight delayed to September 1986, then cancelled)
		Smith	
		Springer	MS
		Garriott	MS
		Nicollier	MS
		Lampton	PS
		Litchenberg	PS
	STS 51K/61A Spacelab D-1	Hartsfield	
		Nagel	
		Buchli	MS/2nd pilot
		Bluford	MS
		Dunbar	MS
		Furrer	PS
		Messerschmidt	PS
		Ockels	PS
	STS 51I	Gibson	Crew reassigned
		Bolden	
		Hawley	MS
		Nelson	MS
		Chang-Diaz	MS
		Konrad	PS
1985	STS 51L	Scobee	
		Smith	
		McNair	MS
		Resnik	MS
		Onizuka	MS
		US Teacher	PS (Added later)

The crew of the STS 51C military shuttle mission. Left to right: Payton, Shriver, Mattingly, Buchli and Onizuka. (*NASA*)

STS 61C	Coats	Crew reassigned
	Blaha	
	Thagard	MS
	Fisher, A.	MS
	Springer	MS
STS 62A	Crippen	
	Gardner, G.	
	Gardner, D.	
	Ross	
	Mullane	
	DoD Manned spaceflight engineer (s)	
STS 61D/Spacelab 4	Brand	Flight delayed,
	Griggs	possible scrub.
	Fabian	MS/2nd pilot (chosen earlier)
	Seddon	MS (chosen earlier)
	Bagian	MS (chosen earlier)
		Plus 2 US PS
STS 61E/Astro 1	McBride	
	Richards	
	Leestma	MS (chosen earlier)
	Parker	MS (chosen earlier)
	Hoffman	MS (chosen earlier)
	Plus 2 astronomers	PS

STS 51J	Bobko	
	Grabe	
	Stewart	
	Hilmers	
STS 51D (16)	**Bobko**	
	Williams	
	Griggs	MS
	Hoffman	MS
	Seddon	MS
	Walker, C.	PS
	Garn	PS
STS 51G	Brandenstein	
	Creighton	
	Lucid	MS
	Fabian	MS
	Nagel	MS
	Baudry	PS
	Al Saud	PS
STS 51I	Engle	
	Covey	
	van Hoften	MS
	Lounge	MS
	Fisher (W)	MS
	Jarvis	PS
	Walker (C)	PS reassigned to 61B
STS 61B	Shaw	
	O'Connor	
	Cleave	MS
	Spring	MS
	Ross	MS
	Mexican	PS
	Walker (C)	PS
STS 61C	Gibson	
	Bolden	
	Chang-Diaz	MS
	Hawley	MS
	Nelson	MS
	RCS	PS
STS 61H	Coats	
	Blaha	
	Fisher (A)	MS
	Thagard	MS
	Springer	MS
	Indonesian	PS
	Wood	PS
STS 61J	McCandless	MS
	Sullivan	MS
	Hawley	MS

Space Shuttle Schedule

Mission	Orbiter	Possible date
18 51G	*Discovery*	17 June 1985
19 51F	*Challenger*	15 July 1985
20 511	*Discovery*	10 Aug. 1985
21 51J	*Atlantis*	26 Sep. 1985
22 61A	*Columbia*	10 Oct. 1985
23 61B	*Challenger*	8 Nov. 1985
24 61C	*Columbia*	20 Dec. 1985
25 51L	*Challenger*	22 Jan. 1986
26 61E	*Columbia*	6 Mar. 1986*
27 62A	*Discovery*	20 Mar. 1986
28 61F	*Challenger*	15 May 1986*
29 61G	*Atlantis*	21 May 1986*
30 61H	*Columbia*	23 June 1986
31 61J	*Atlantis*	8 Aug. 1986

*These three missions *must* be launched near to these dates. They are the Halleys Comet Astro 1 observation mission and missions to deploy the Ulysses solar polar orbiter and Galileo to Jupiter.

Explanation of numbering system

NASA's 'year' runs from 1 October to 30 September. So, 61A means 1986 (actually 9 October 1985), Cape Canaveral, first mission. 62A means 1986, Vandenberg, first mission. 61G means the seventh mission from the Cape in the 1986 fiscal year. Even if flights are scrubbed and the schedule changes these numbers apply.

POST SCRIPT

Latest Flights 1985

Flight 106
Launched 6 June 1985
Craft Soyuz T13
Crew Vladimir Dzanibekov Cdr
Viktor Savinykh, Eng

58th Soviet manned space flight. Savinykh 68th person to make two flights. Dzanibekov 2nd person to make five flights (1st Russian and youngest).

Flight 107
STS 51G/Discovery

Launched 17 June, 1985
Duration 7 days, 1hr 41min
Crew Brandenstein, 42, (2nd Flight)
Creighton, 42
Lucid, 42
Nagel, 38
Fabian, 46 (2nd Flight)
Baudry, 39
Al-Saud, 28

France the first country to fly both US and USSR missions
First manned spaceflight to carry two foreigners: Baudry, Al Saud.

Lucid oldest woman in space at 42.

Al-Saud youngest non-US-non-USSR space person

STS 51F/Challenger

Launch Due 12 July, 1985
Duration 7 days
Crew Fullerton, 48 (2nd Flight)
Bridges, 41
Musgrave, 49 (2nd Flight)
Henize, 58
England, 43
Acton, ?,
Bartoe, ?,

Henize oldest man in space at 58

STS 51I/Discovery
Launch Due 20 August, 1985
Duration 7 days
Crew Eagle, 52 (2nd Flight)
Covey, 39
Lounge, 39
Fisher, W, 39
Van Hoften, 41 (2nd Flight)
Jarvis, 40

Possible EVAs by Fisher and Van Hoften.

STS 51J/Atlantis
Launch Due 26 September 1985
Duration Unknown (military flight)
Crew Bobko, 47 (3rd Flight)
Grabe, 40
Hilmers, 35
Stewart, 43 (2nd Flight)

New Assignments
STS 61B Mexican PS Rudolfo Neri Vela

STS 61F
Crew Hauck, Cdr
Bridges Pil
Hilmers MS
Lounge MS

STS 61G
Crew Walker D., Cdr
Grabe Pil
Fabian MS
van Hoften MS

New Astronaut Group
11th Group of NASA candidate astronauts chosen in July: 13 people including 6 pilots and 7 mission specialists, 2 of whom are females.

Glossary

Explanation of major abbreviations in this book

ALT	Approach and Landing Tests (Shuttle)
ASCS	Automatic Stabilization Control System
ASTP	Apollo Soyuz Test Project
ATDA	Augmented Target Docking Adapter (Gemini)
CDR	Commander
CM	Command Module (Apollo)
CMP	Command Module Pilot
CNES	Centre National D'Etudes Spatiales
CSM	Command Service Module (Apollo)
DOD	Department of Defense (US)
DM	Docking Module
ELSS	Environmental Life Support System
EO	Earth Orbit
EO/D	Earth Orbit/Docking
ET	External Tank (Shuttle)
EVA	Extravehicular Activity (space walking)
G	Acceleration force
GATV	Gemini Agena Target Vehicle
HHMU	Hand Held Manoeuvring Unit
ICBM	Intercontinental Ballistic Missile
IRBM	Intermediate Range Ballistic Missile
IUS	Inertial Upper Stage (Shuttle Stage)
LEVA	Lunar EVA
LFB	Lunar Fly-By
LL	Lunar Landing
LM	Lunar Module (Apollo)
LMP	Lunar Module Pilot
LO	Lunar Orbit
LRV	Lunar Roving Vehicle
MA	Mercury Atlas
MDA	Multiple Docking Adapter
MET	Modular Equipment Transporter
MMU	Manned Manoeuvring Unit (Shuttle)
MOL	Manned Orbital Laboratory
MR	Mercury Redstone
MS	Mission Specialist (Shuttle Astronaut)
MSFE	Manned Space Flight Engineer (Shuttle DOD Passenger)
NASA	National Aeronautics and Space Administration
OFT	Orbital Flight Test (Shuttle)
OMS	Orbital Manoeuvring System (Shuttle)
P	Pilot
PFTA	Payload Flight Test Article
PL	Pilot
PLSS	Portable Life Support System
PS	Payload Specialist (Shuttle Passenger)
RCS	Reaction Control System
RMS	Remote Manipulation System
RP1	Rocket fuel
RSCS	Reaction Stabilization Control System
SEVA	Stand-up EVA
SIM	Space Instrument Bay (Apollo)
SIVB	Saturn Rocket Stage
SLC	Shuttle Launch Complex
SM	Service Module (Apollo)
SP	Senior Pilot (Apollo)
SPAS	Shuttle free-flying platform
SRB	Solid Rocket Booster (Shuttle)
SSME	Space Shuttle Main Engine
STS	Space Transportation System (Shuttle)
TEEVA	Trans-Earth EVA
TPS	Thermal Protection System
USA	United States of America
USA	US Army
USAF	US Air Force
USMC	US Marine Corps
USN	US Navy
USSR	Soviet Union

Metric and Imperial Units and Conversions (* = exact)

Column One	*Column Two*	*To convert Col. 1 to Col. 2 Multiply by*
Length		
inch (in)	centimetre (cm)	2.54*
foot (ft)	metre	0.3048*
yard (yd)	metre	0.9144*
mile	kilometre (km)	1·609344*
Area		
square inch	square centimetre	6·4516*
square foot	square metre	0·092903*
square yard	square metre	0·836127*
acre	hectare (ha) (10^4 m^2)	0·404686*
square mile	square kilometre	2·589988*
Volume		
cubic inch	cubic centimetre	16·3871*
cubic foot	cubic metre	0·028317*
cubic yard	cubic metre	0·764555*
Velocity		
feet per second (ft/s)	metres per second	0·3048
miles per hour (mph)	kilometres per hour	1·609344
Acceleration		
foot per second per second (ft/s^2)	metres per second per second (m/s^2)	0·3048*
ounce	gram	28·349523125
pound	kilogram	0·45359237*

Index

A.

N.

O.

P.

R.

S.